Elusive Saviours

“The industrial world’s environmental debts are enormous, go back a long way, and still go mostly unaccounted for.”

Aubry Meijer, Global Commons Institute. “Green Rights for All: The Earth View” *In* ECN *Environmental Review*, July 1992.

Hans Heerings and Ineke Zeldenrust

ELUSIVE SAVIOURS

Transnational corporations and sustainable development

Translated by Lin Pugh
Edited by Niala Maharaj

International Books

CIP-GEGEVENS KONINKLIJKE BIBLIOTHEEK, THE HAGUE

Heerings, Hans

Elusive Saviours: transnational corporations and sustainable development/Hans Heerings. – Utrecht : International Books. -Ill.
Published in cooperation with SOMO – With Bibliography
ISBN 90-6224-978-7
NUGI 661/684
Key words: transnational corporations and environment policy

We would like to thank Novib for financially supporting this publication. Novib supports non-governmental organisations in the South working to assist underprivileged peoples. Novib's objective is to contribute toward sustainable development by means of structural poverty alleviation. It believes that the information in this book can help Novib and its partners in the South form a balanced opinion and strategy on this.

Cover design: Marjo Starink
Printing: Bariet

International Books
A. Numankade 17, 3572 KP Utrecht
tel. 31 (0)30-2731840, fax. 31 (0)30-2733614

Preface

Since the seventies, transnational corporations have been attacked for their role in the world's environmental problems, by critics both at home and in the South. Justifiably so, and not only because of the pollution caused by their factories. The major role they play in the international economy and international political relations also has an effect on the environment.

Transnational corporations are one of the most important players in the development of economies as a whole, in the globalization of these economies, in the development of employment worldwide, in exploiting the cheap labour in the South and pushing labour productivity in the North beyond the bounds of reasonable. Transnational corporations shape modern consumerism in the North and increasingly also in the South. They more and more dominate in areas of technology, protecting their innovations and developing whichever ventures will be profitable under current market circumstances.

As transnationals have become more powerful, politicians have beome more adrift. Transnationals are in control of the international agenda. Instead of organizing on the principles of justice, equal opportunities and sustainable development, politicians discuss free trade and deregulation: the survival of the fittest.

The corporations are not, however, laying on a bed of roses. They are never able to enjoy a total monopoly, and are forced to compete for the preference of consumers and decision-makers. The growing environmental awareness of the public has forced these giants to look into their environmental performance. They prefer doing it themselves, rather than having others do it for them.

The image they are building is green. It's not all vile tactics; many people working within transnationals believe that good environmental performance is a matter of social responsibility. These people were greatly supported by the "sustainable development" wave that hit the companies before and after UNCED in 1992. If we look solely at environmental pollution and resource use per unit of production, we can also conclude that they have made progress.

Friends of the Earth International (FoEI) does not see transnational corporations as "enemies". FoEI is actively encouraging cooperation with the companies, within the framework of its Sustainable Societies Programme. Transnationals can lead the way in product and process modernization.

They will not, however, be the leaders in the far-reaching reduction of the use of mineral resources, energy and land by industrial societies that FoEI advocates. This is against their interests. Governments need to act in this sphere. That is why governments, as this study shows, can no longer avoid coming into conflict with transnational corporations.

John Hontelez, chair FoEI
Ricardo Navarro, vice-chair FoEI, winner of the 1995 Goldman Environmental Prize

Table of contents

CHAPTER I

Transnationals as saviours of the environment?

During the preparations for UNCED and the huge environmental conference itself, government representatives were unable to reach clear and binding agreements on essential measures needed to solve global environmental problems. There were too many differences in the positions of Northern and Southern governments, so more and more politicians began to look to the corporate community as the saviour. And international corporations chose UNCED as the vehicle by which they could present themselves as 'the green alternative'.

With their tremendous economic power, these corporations have enormous influence on the environment. This puts them in an undeniable position to stimulate and develop sustainable development. However, they are simultaneously a source of local and international environmental damage, along with national industries, agricultural companies and consumers.

But this raises a question: can the international corporate community, one of the perpetrators of the world's environmental problems, successfully become the saviour of the environment? Should we all follow this new international guide with its new green coat, now that international politics has come to a standstill? Can transnational corporations offer a realistic opportunity for world-wide sustainable development? Is their omnipotent economic power an obstacle or a bridge to sustainable development? These are the central questions of this book.

The first chapter discusses how the difference in positions between the North and the South hampered agreement at UNCED on binding international environmental measures and environmental control. At the end of the chapter we will see how international corporate life lobbied intensively for this conference decision. The most important argument used by the corporate lobby was that the transnational corporate community is able to draw up an effective environmental policy independent of the political community. Mandatory international laws would, they say, only hinder corporate efforts to attain sustainable development.

The following chapters discuss whether the corporations will in fact use this freedom in the interests of the environment. They look at a number of specific characteristics of transnationals that are important in their relation to the environment:

- the degree of participation of transnational corporations in environmentally sensitive branches of industry
- their vital position as a link between world-wide mining, agricultural, production, trade, distribution and service activities
- their mobility and freedom in choosing the locations for investments
- their dominant position in the field of technological application and development
- their elusiveness in relation to international laws and regulations.

The final chapter discusses the far-reaching consequences of the new GATT and WTO agreements for national environmental policies, and what this implies if international environmental management is left in the hands of transnational corporations.

CHAPTER 2

The North and the South in the global environmental debate

Selective awareness of the global environmental problem in the North

Since the industrial disasters in Bhopal (Union Carbide), Basel (Sandoz) and the nuclear power plants in Chernobyl, the rich countries of the North have experienced a massive growth in awareness of the risks of modern chemical technology. Most attention has been focused on the public health risks involved in the chemical industry. This is an industry which has developed on a large-scale right in the middle of, or close to, urban areas. Governments have translated this increased awareness into regulations and legislation concerning disaster plans, risk analyses and safety measures. Take for example the European Community's post-Seveso guideline:

The Seveso guideline. Example of a binding regulation

The Seveso guideline was effected in the EC in 1984, and requires that member states ensure that companies calculate risks, develop a disaster plan and actively inform the local population. Translated into the Dutch context: 74 Dutch chemical industries are now obliged to inform local people of the external risks of their activities. They had to draw up an External Safety Report which included a description of the production process, and an analysis of the risks.

Besides the risk of industrial disasters, the 1980s saw an increase in awareness of the problems of acid rain, the greenhouse effect and the consequences of the hole in the ozone layer. Renewed attention focused on the threat of extinction to many species of animals and plants and the value of "biodiversity". It became generally accepted that not only the immediate, but also the international environment was threatened. For the first time, people seriously looked at what was happening beyond their own borders and discovered that tropical rainforests – the lungs of the world – were at risk of extinction. They discovered that the levels of CO_2 emissions were so high that the world had been transformed into a greenhouse, and realized that CFCs were responsible for an expanding hole in the ozone layer. People realized – as the Club of Rome and Meadows had already discovered – that there are limits not only to our raw materials, but also to other natural resources: soil, air, water, flora and fauna.

The optimistic theory of sustainable development

The gravity of the global environmental threat threw up solutions which can be summed up in the catchphrase "sustainable development," which, since the publication of the United Nations' World Commission on Environment and Development report, Our Common Future (also known as the Brundtland Report) has been assimilated into the North's philosophy on the relationship between environment and economy. At the base of this concept lies the notion that environmental conservation and economic growth can be combined, and that ecological considerations should become integrated into all political and economic decisions. Regarding the environment, its use, reparation and maintenance, as "natural" capital, became viewed as not necessarily in conflict with the aims of economic development. The North assumes that this growth-optimistic vision can be applied world-wide.

Sustainable development?

The term "sustainable development" was first introduced in 1980 in the "World Conservation Strategy", a plan created by the IUCN and the World Wildlife Fund. It only came into wide usage, however, after publication of the so-called Brundtland Report, which defined it as: "development which meets the needs of the present without compromising the ability of future generations to meet their own needs".

Sustainable means no reductions to the environmental capital:

- that the extent of the worlds available resources (for example the amount of land available for agriculture) is not reduced,
- that the system of ecological regeneration remains intact (for example that agricultural land is given sufficient time to lay fallow),
- and the biological diversity and resilience of ecosystems is respected."

Theoretically, this implies an end to the dichotomy between economic growth and a clean (or cleaner) environment. No longer can negative environmental effects be regarded as irrelevant. Only sustainable/clean growth is acceptable growth.

Sustainable development has become a container term. A World Bank report noted 60 different definitions. Ecologists describe it as the: "harmonious relationship between the population and nature, the conservation of the integrity of ecology and a humane existence."

Others see "sustainable" in the light of present production and consumption. What, then, are the possibilities for keeping this intact and/or continuing growth? Critics, particularly from the South, argue that current activities described as "sustainable" are aimed, instead, at growth – or serve that purpose – and will remain so as long as the problem of poverty and unequal distribution remains.

Despite the diversity in definitions, there is a reasonable degree of agreement as to the necessary conditions for embarking on "sustainable development":

- the development and introduction of production techniques and policy instruments which reduce pollution
- the development and introduction of techniques for recycling wastes and non-renewable products (closing cycles)
- severe limitations on the total use of non-renewable resources and the

use of renewable resources (on condition they continue to be regenerated)
- the reduction of polluting and "natural resource-intensive" consumption patterns.

It seems unlikely that a start will be made in applying these conditions (particularly the third and fourth ones) in the short term.

Global environmental problems and consumption patterns

It is not difficult to understand the popularity of the theory of sustainable development in the North. While it has, on the one hand, precipitated a revolution in ideas on patterns of economic progress, it also confirms that, in its patterns of consumption, the highly industrialized world is addicted to growth. The theory recognizes that environmental measures are needed to create a sustainable economy, and, at the same time, promises that this can be achieved without loss of wealth or welfare.

It sounds too good to be true. It is a false perspective. This is clearly illustrated by the practice of world energy management. If we look closer at our energy supply – *the* best measurement of economic growth and welfare – we see that it is not possible to maintain or even reduce CO_2 emissions in world energy production while aspiring, at the same time, to provide every citizen in the world with the same amount of energy per capita as the average north American citizen.

The unequal responsibility for the greenhouse effect

The three hundred scientists from forty different countries that comprise the "Intergovernmental Panel on Climate Change" (IPCC) wrote in their report "Climate Change: The IPCC Scientific Assessment" that the only way to avert an ecologically disastrous warming of the earth is to drastically reduce carbon dioxide emissions (60% relative to the 1990 level). In September 1994 the IPCC meeting in Maastricht reaffirmed this view.[1]

The greenhouse problem is directly caused by emissions of a number of gases including CO_2, the bulk of which are due to the large quantity of emissions in highly industrialized countries. The so-called "stock effect" is important in looking at the problem of CO_2 emissions. This term refers to the amount of pollution the earth can cope with before negative environmental effects are activated, a threshold that has long been crossed, so that, at present, reductions in CO_2 and SO_2 emissions will not immediately lead to less environmental pollution and the reversal of the greenhouse effect. In general, drastic reductions of 70 to 95 per cent in emissions are needed to combat environmental problems caused by the present stock of polluting substances.[2] The scale on which change is needed means that whole sectors will have to change dramatically before any improvement will be seen.

Estimated cumulative industrial CO_2 output 1960-1989 and 1980-1989[3]
Ranged according to area, in giga tons carbon and percentages

	A 1960/89	D %	C 1980/89	D % C of A
Far East	6903	3	3077	45
Centrally controlled economies Asia	12114	6	5385	44
Middle East	3586	2	1538	43
Africa	4520	2	1538	34
Eastern Europe	47184	23	15385	33
Oceania and Japan	9932	5	3077	31
Other American	7697	4	2308	30
Western Europe	41926	20	7692	18
United States	67671	33	12308	18
North America excluding USA	4877	2	769	16
Total	206410	100	53077	

The Northern and the Southern agenda for the 21st century

Armed with the theory of sustainable development, the world community met at the United Nations Conference on Environment and Development in 1992 to draw up a political agenda for the 21st century to end non-sustainable development through a series of wide-ranging measures. During the preparations, it had become painfully apparent that the agenda of the North was very different to that of the South, in that its starting point was not redistribution and equality, either in relation to people or the environment. In the words of Anil Agarwal, director of the Centre for Science and Environment, New Delhi:

> "No discussion took place – in Brazil – on environment and aid, environment and trade, environment and poverty, environment and wealth and environment and patterns of consumption."[4]

While non-governmental organizations were deeply involved in the preparations, the final meeting was a meeting of states, the agencies responsible for fulfilling the agreements.

According to resolution 44/228 of the United Nations, the conference was meant to deal with:

- Protection of the atmosphere by combating climate change, the destruction of the ozone layer and cross-border air pollution
- Protection of the quality and supply of freshwater resources
- Protection of the oceans and all types of seas
- Protection and management of land supplies by combating deforestation and desertification
- Conservation of biological diversity
- Environmentally sound management of biotechnology
- Environmentally sound management of wastes, in particular toxic wastes

At the preparatory meetings, the following "cross-sectoral issues", or themes, relating to all of the above points, were added to the list. Countries in the South, in particular, emphasized that in order to come to an integrated approach, these themes had to be discussed:

- Patterns of production and consumption

- Access to and transference of environmentally sound technology
- Mechanisms of financing
- Legal aspects
- "Human Resource Development", in particular education and training
- Economic instruments
- Guaranteed sustainable food supply
- Institutional facilities, and new structures in order to integrate environmental and economic development into decision making processes
- The improvement of the living and working conditions of the poor in urban and rural districts by dispersing with poverty and improving the quality of life.

The goal of UNCED was to agree to an "Earth Charter", a sort of declaration of the rights of the environment, draw up an action plan for the future, called Agenda 21, and finalise two conventions, a on world-wide climate change (the greenhouse effect) and biodiversity.

Money in exchange for a clean environment

At the UNCED preparatory meetings, it became only too clear that the Third World's environmental problems are closely entwined with the old problems of poverty and wealth. Much of the poverty in the South and the resulting environmental problems are the result of the inequitable workings of the economic system devised by the industrialized countries of the North.

> "As recent reports of the World Bank, the United Nations Development programme and others show, the poor developing countries now actually measurably subsidize the rich countries through structural adjustment and regimes of export-led growth, inequitable barriers to trade, low commodity prices and the now seemingly endless regime of debt repayments. All this adds up to a considerable net resources transfer from South to North."[5]

From the perspective of the underdeveloped and poor countries of the South, the foremost environmental problems are the major, everyday ecological problems caused by underdevelopment and poverty, such as: the lack of clean drinking water, infertile soil as a result of erosion, salinification, the decreasing level of the water table, immense air pollution in the cities and the mounds of toxic waste in their back yards. In this part of the world, it is a luxury to include ecological considerations in political-economic decision-making processes. They hold the North as first and foremost responsible for global problems such as ocean pollution and the destruction of the ozone layer. Niala Maharaj summed up this view as follows:

> "It is your skin cancer. Solve the problem yourself. Our problems are the measles, whooping cough and even bubonic plague." (*De Volkskrant*, 9 November 1991)

Despite its raised consciousness of the global environmental problem, the North still shows no interest in the everyday problems of the Third World. The North still regards these as "their" problems. For centuries, the North has treated nature as a useful resource. Only recently has it begun to look at "Nature" as a communal area, the so-called "global commons". Attention is given to the pollution of the oceans, the atmosphere, space and the poles, as well as to the destruction of our tropical rainforests and other areas rich in biodiversity. People in the North have recently begun to regard these areas which are essential in the natural regeneration of the global ecosystem as "communal heritage for humanity". They are less interested in local environmental problems. Political discussions on solutions therefore tend to concentrate on the global management of these areas.

Most of the commons, however, are situated on nationally controlled territory in countries in the South. Understandably, these countries regard any Northern involvement in the management of these areas as illicit meddling in home affairs.

In a declaration published in Crosscurrents, 38 NGOs from 25 countries stated that they: "were concerned that the introduction of the concepts of 'global commons' and 'communal heritage for humanity', if they do not include the protection of the rights of rural populations, would lead to an increase in the amount of control exercised by the North and particularly by transnational corporations over the natural resources of the South."[6]

It is not surprising, against the background of economic inequality and poverty, that governments in the South demand that the North finances the extensive, world-wide environmental measures directed toward the conservation of the commons.

During the preparatory negotiations for UNCED, Malaysian government representatives stated bluntly: "If you do not want to discuss the development demands, we will not discuss the environment. If we die, we will all die."[7]

Environmental degradation hits hardest in the South

There are several reasons why environmental degradation hits hardest in the South:

- Due to the burden of short-term problems, environmental problems and particularly possible long-term environmental consequences receive less government attention.
- Environmental legislation in the South is generally not as advanced as in the North and lacks in particular institutional support. There are, in other words, too few organizations and services to control and carry out the law.
- One of the consequences of the poverty and inequitable distribution of wealth characteristic of Third World countries is the absence of effective environmental measures. The result can be disastrous. Many people cannot read or write and cannot therefore read product instructions and safety instructions. They do not have the tools to be able to inform themselves of the possible dangers and problems, even if this information is provided. Employees are often not adequately trained to understand the dangers of work which is possibly dangerous to the environment or their own health.

Another aspect of poverty in Third World countries is the number of children working in factories and mines. Children are much less aware of dangerous situations in their labour conditions, and they certainly are unable to read the safety instructions.

Health and health care, certainly in the poorer sections of the population, is inadequate. As a result, peoples' health is more susceptible to the effects of environmental pollution, and victims have less access to care.

Poverty also means that in many countries there are no financial reserves for actually providing environmental protection. Drainage systems in many areas are, for example, either badly constructed or non-existent. The water which people rely on for their everyday use is unclean. Toxic goods often have to be transported over badly kept roads, with the unsurprising result of serious mishap.

Scarcity of finance in the South seriously hinders research into solutions to environmental problems, so there are fewer barriers to the establishment of certain (polluting) production activities than in the North.

- Generally speaking, there are fewer organizations raising public awareness of environmental questions in the Third World. The Chipko movement in India (Save the Trees) and the large opposition movements to factories which produce radioactive wastes in Thailand and Malaysia are among the best known exceptions, but they often work under more difficult conditions than their counterparts in the North. Most work at the grassroots level and most have, as yet, little political say. This is largely a result of the fact that, at least in the short term, most of the problems mainly effect the poor.
- A final reason for the greater susceptibility of Third World countries to environmental pollution is the fragility of the natural balance there. Mercury discharges resulting from gold panning in the Amazon rainforest have been disastrous. Insecticides threaten to poison huge areas of desert.

Is there life after UNCED?

We know what happened at UNCED. According to the Climate Agreement, emissions of greenhouse gases will be reduced to 1990 levels by the year 2000. No agreement was reached on specific reduction aims for decreases in the emissions of carbon dioxide (CO_2) and the other gases. That is a step backwards from the 1988 Toronto climate conference, where a 20 per cent reduction in CO_2 emissions relative to a base year to be determined in 2005 was agreed. The developed and underdeveloped world were not able to agree on a definition of the prase "joint implementation of the convention".

It wasn't until April 22, 1994, that U.S. President Bill Clinton signed the Biodiversity Convention and presented it to the House of Representatives for ratification. The Bush government had refused to sign it during UNCED, arguing that it would restrain biotechnological development and undermine intellectual property rights.

The Biodiversity Convention

One of the most important issues in the discussions on the Biodiversity Convention was the distribution of the products of biodiversity. Genetic information present in plants and animals can be invaluable to commercial and non-commercial biotechnological research. At the present time, however, no mechanism exists for returning the profits made with the help of this information to the countries and population groups which are responsible for the conservation of the plants and animals concerned. Unfortunately, the negotiators were not able to agree to effective, mandatory solutions. The agreement was only able to state that parties have the right to demand that genetic material be exported under permit, and that parties can demand a share in the profits, either in the form of financial remuneration or in (bio)technological cooperation.

The Bush administration openly spoke of protecting corporate interests. As "intellectual property rights" was regulated to the satisfaction of the international (Northern) business world in the GATT agreement of April 13, 1994, this argument was no longer relevant to the Clinton administration. The Bush administration also argued that the proposed measures would cost money which the US preferred to spend on combating domestic unemployment. While some funding became available for the protection of rainforests, the proposed "Rainforest Convention" never got off the ground. In its place came the non-binding "Forest Declaration". As well-worded as the 600-plus page "Agenda 21" is, unlike other conventions it does not commit its signatories to any firm course of action and cannot live up to its subtitle: a "blueprint for *action*".

The slow implementation of the UNCED agreements

The process of implementing the watered down UNCED conventions has been slow. In November 1992 the General Meeting of the United Nations established the Commission for Sustainable Development (CSD) whose first substantial meeting took place in June 1993, when 53 member-states discussed the Secretary General's proposal for monitoring the progress of implementation of Rio agreements and set a deadline for its work, 1996.

Developing a financing mechanism for the implementation of the UNCED conventions in developing countries turned out to be a laborious process. After 16 months of negotiations between representatives of more than 80 countries, an agreement was reached and signed on March 16, 1994 in Geneva. The Commission appropriated the already existing Global Environmental Facility (GEF) as the interim financing mechanism for the UNCED Conventions relating to biodiversity and climate change. Diverse Northern countries agreed to donate a total of to billion dollars for this purpose between 1994 and 1996. While this is not much money in relation to the size of the global environmental problem, it is already clear that it will be difficult to find development projects which fulfil the criterium of "sustainable development" and profit from this funding. See appendix 2 for these and other critical notes on the GEF.

By January 1994, of the 166 countries that had signed the climate convention, only 53 countries had ratified it (a minimum of 50 was needed to set the convention in motion). These countries are eligible for participation in the first "Conference of Parties to the Climate Convention" in Berlin, which only took place in 1995 because of the slow pace of the ratification process. And ratification is only the first step in implementing the convention. Adaptation to national rules and regulations is the next step. It is not always clear what that should involve as there are many vagaries and very few aspects are actually discussed in the convention. Meanwhile an intergovernmental commission began negotiations for the 1995 conference. Representatives of non-governmental organizations campaigned for the acceptance of the 20 per cent CO_2 emission reduction goal set by the Toronto conference in 1988 and try to gain clarity on the concept of "Joint Implementation". They wanted to establish the level of participation of both developed countries and underdeveloped countries in the reduction goals, as well as the position of the Newly Industrialized Countries. The Conference also discussed establishing its own financing mechanism and a system of observation. In the end, the Berlin Climate Conference did not succeed in establishing reduction goals.

The transnational corporation as redeeming angel

The incapacities of the world politic and the tremendous differences in positions between the North and the South has caused more and more governments and politicians to turn to the transnational corporate community for assistance.

But what can the transnational corporations do if politics fail? Are they the angels of environmental redemption? Can they fulfil this role? The corporate viewpoint was clear and totally in agreement with the results of UNCED, namely, that *binding global environmental measures for international business be avoided.* With the help of an extensive public relations campaign, top managers of the world's largest corporations informed UNCED participants and the general public that they had *voluntarily* chosen to embark on the road to "sustainability". To do this they need economic growth, free trade and open markets. Profit making *and* environmental protection can be combined, they argue, in a system in which the damage to natural resources is compensated in the cost price of goods, and in which nature is patented. According to the influential lobbying organization, the Business Council for Sustainable Development, this is the road to a sustainable future.

As Indian ecologist, Vandana Shiva put it, "Governments have distanced themselves from their responsibilities. As of now, it is a fight between transnationals and citizens."

The international business community's successful lobby during UNCED

The choice for Maurice Strong – a Canadian businessman and multi-millionaire – as Secretary General of the UNCED was an early sign of the influence of the international business community on this conference. In the early preparatory stages, Strong appointed the Swiss captain of industry, Stephen Schmidheiny, to be the most important advisor for business and industry, and he approached 48 top managers, of corporations including Dupont, Shell, Dow Chemical, Ciba-Geigny and Mitshubishi, to form the Business Council for Sustainable Development (BCSD), which published a book, *Changing Course* before UNCED began, outlining their proposals.

As UNCED opened, Strong told Schmidheiny in front of scores of reporters, "No contribution [to UNCED] has been more important than yours (i.e. BCSD)."[8]

He also remarked that he held the International Chamber of Commerce (ICC), a major lobby against binding environmental measures, in "high esteem". Approximately half of the corporations in the BCSD were represented on the ICC's board of directors, a board which was partly responsible for the failure of the detailed proposals presented by Norway and Sweden, advocating binding environmental measures, at the preparatory meeting of UNCED in New York. At the end, the final concept of Agenda 21 did not include any proposals controlling the multinationals, but stressed rather the role of the business community and industry in environmental protection and the importance of the voluntary measures they had already taken.

In a perfect division of roles, the task of the BCSD was to show the public the environmentally friendly face of the international corporate community, and even to promote calculating the environmental costs in the price of goods. Meanwhile the ICC surreptitiously lobbied with all its might against inclusion of measures in Agenda 21 which would make this possible.

The international business community also set up lobbying organizations on various subject areas to keep a finger on the UNCED pulse. An example is the Global Climate Coalition (GCC) which worked diligently on influencing politicians who were in favour off reductions in carbon dioxide emissions. As a result, UNCED failed to develop a Climate Convention obliging signatory countries to reduce carbon dioxide emissions.

International corporations such as Ashai Glass, Atlantic Richfield, ICI, Swatch and 3M also influenced UNCED in indirect and subtle ways. These corporations were the most important donors to the Ecofund, a non-profit organization working from Washington DC, set up to help finance the UNCED. Global Forum, the alternative environmental conference for non-governmental organizations, also received money from the business community.

By the time UNCED drew to a close, functionaries of transnational corporations breathed sighs of relief. They could now embark on a period of free-market international environmental protection without the involvement of national governments in their business activities. With little hindrance from environmentalists, they were now free and well equipped to continue working through their full agenda load for the GATT negotiations (on the liberalization of world trade).

Spurred on by its own successes, the Business Council for Sustainable Development, originally created as a one-off coalition of 48 captains of industry for the duration of the UNCED process, decided to continue operations. It now acts as the "conscience of the international business community". Hugh Faulkner, president director of the BCSD, commented: "We will be working with other groups and sectors to try and promote the change in policy, in corporate governance and various areas which are preconditions to sustainable development."[9] He referred specifically to the GATT when he said: "We have to make clear to the world and the politicians that there is a higher order of public interest in this point."

Before we discuss the actual importance of GATT for the environment, and transnational interests within this, the following chapters will firstly analyse the involvement and the role of transnational corporations in the creation of and solutions to the world's environmental problems.

Summary

With their optimistic theories of sustainable development, Northern politicians argue that it is possible to develop in a sustainable way without risking wealth and welfare, in other words, without fundamental changes in patterns of consumption and production. The North focuses its attention on the problems of the global commons, such as tropical rainforests, which it views as the common heritage of humanity, essential in the natural regeneration of the global ecosystem. Political discussions on the solutions to global environmental problems are directed towards possible forms of global management of these areas.

During UNCED it became increasingly apparent that this was a false perspective. The primary environmental problem from the perspective of underdeveloped and poor countries in the South is the tremendous, everyday ecological problem which is linked to underdevelopment and poverty. This takes a variety of forms: lack of clean drinking water; infertile soil due to erosion, salinification and receding water tables; air pollution in the cities and piles of toxic waste surrounding human habitats. In political-economic decision-making processes, ecological considerations are luxury items.

According to the South, the North is primarily responsible for such world-wide problems as the pollution of the ocean and the destruction of the ozone layer. The "commons" are mostly found within the borders of countries of the South, and these countries resist the North's interference as unlawful meddling in or on their national territory.

These differences in position between the North and the South led to a situation where more and more governments and politicians looked to the international business community rather than governments as the redeemer. The BCSD, a one-off coalition of 48 captains of large transnational corporations, lobbied intensively to achieve this, and by the time UNCED came to a close, the transnational corporations were able to draw deep sighs of relief in the knowledge that they could embark on an era of free-market international environmental protection without the interference of international governing bodies.

CHAPTER 3

Involvement in the international environmental problem

Prominent presence in environmentally sensitive sectors

Transnational corporations dominate and influence the larger percentage of total world production in many branches of industry. But their power is not visible in the statistics, as they are not listed as a separate category and exact data is scarce. Still, they "influence at least a quarter of all production activities in the world, they are responsible for 70 per cent of the international trade in goods and 80 per cent of the total amount of land cultivated for export crops."[10]

Dominant industries in sensitive sectors

CFCs, one of the most important contributors to the hole in the ozone layer, were almost exclusively produced by large chemical concerns – most of which are transnational corporations. E.I. Dupont de Nemours and Co. alone was, until very recently, responsible for 25 per cent of the total world production of CFCs. The Treaty of Montreal agreed to slowly bring their production to a halt. (See also Chapter 7 of this book.)

The pesticide industry too is strongly concentrated: The 20 largest pesticide corporations control 94 per cent of the world market. There is also a high concentration of transnationals producing the various sub-groups of pesticides. In the aluminium industry, the 20 largest corporations control 90 per cent of the – highly polluting – world bauxite production.

Perhaps more important than the omnipotence of the transnational corporations is their active involvement in environmentally sensitive areas of industry, including mining, the chemical industry, heavy metals, wood/paper, agro-business and the petroleum industry. And much of the production is concentrated within a limited number of corporations.

Consequently, transnational corporations are also responsible for a large number of environmental problems and in a way to those caused by activities related to their products.

The share of transnational corporations in the total production (through human activities) of "greenhouse gases". *[11]

CO_2: Approximately half of all carbon dioxide emissions caused by human activity
(The total of carbon dioxide emissions resulting from human activity: car emissions, OECD oil and gas use, half the OECD use of coal and half of the use of fossil fuels in developing countries)

Methane: Between 10 and 20 per cent of the methane emissions resulting from human activity
(The total of methane emissions resulting from human activity: half of the emissions in oil and gas production, and half of the emissions from coal mines).

CFCs: Approximately two-thirds
(The total includes aerosol sprays, air-conditioners, solvents, and applications in refrigerators in the OECD).

Other: Approximately half of the other gases, such as nitrous oxide and ozone
(the anthropogenic total corresponds roughly with the same categories of fossil fuel as by CO_2).

---

* We refer here to emissions which transnational corporations can influence (through their own or governmental measures).

Before we can estimate the degree of involvement, responsibility and liability of transnational corporations, we have to define what we mean by "transnational corporations", and what we define as "being influenced" by them.

How far do involvement, responsibility and liability go?

The question of how to define transnational corporations may seem purely academic, but it is not. Our definition determines how we measure the involvement and liability of transnational corporations in environmental problems and, as shall be seen in other chapters, how responsibile they are for environmental problems and what sorts of measures they should take to solve them.

Transnational corporations are often defined in terms of ownership: the transnational corporation as a concern with production and sales units in two or more countries. In this definition, because it owns the subsidiary, the corporation as a whole is responsible for the environmental effects of the subsidiary.

This definition is inadequate. In general the relationship is more than simply one of owner/subsidiary. The subsidiary is usually, in many ways, a part of the larger corporate whole. At the very least the subsidiary has a financial relationship and as well as a general policy relationship with the owner company and, depending on the specific structure of the corporation, the transnational activities of the subsidiaries are tuned, to a greater or lesser degree, to each other. Some transnational corporations even have complete vertical integration; all stages in the production chain – often world-wide – are centrally organized, from mining to production and trade. The more integrated the activities are within the corporation, the greater is the corporation's responsibility for the activities of its subsidiaries. Investment decisions which are taken in one branch of the corporation have, after all, repercussions for all the other sectors of the corporation.

Liability is an important aspect of the relationship between the parent company and its subsidiaries. Most countries grant businesses a "limited liability" status. Under limited liability, shareholders are not personally liable for company debts, and the parent company is liable for the *contractual obligations* of the subsidiary only if the subsidiary is a fully integrated part of the corporation. OECD member states recognize the liability of the parent company for the obligations of its subsidiaries. They do not recognize the corporation's *general liability* for environmental or other damage caused by its subsidiaries. There are many other forms of responsibility in which liability should be recognized. The parent company should be seen as liable for the damage caused by its subsidiary if, for example, it did not exercise sufficient control over the subsidiary or did not supervise it adequately, if it did not warn it of known risks, or supplied the subsidiary with defective parts or products.

An equally prevalent but wider definition of transnational corporations looks at who controls what. This fits in more closely with the present trend of less direct ownership and more contracting out of segments of production, using a network of businesses for various forms of "cooperation". The transnational corporation is seen as a central organization controlling production in a number of countries. According to this definition, transnational corporations are characterized not by the relationship of ownership, but by control over production. They control numerous business activities without actually investing in them. Such companies are practically, but not formally, subsidiaries. If the transnational uses nationally operating subcontractors, these also become, according to this definition, part of the "transnational". The effect of the subcontractors' activities on the environment bear "per definition" on the environmental practice of the transnational as a whole.

If we define transnationals in this way, the question of whether national companies cause less pollution than transnational corporations becomes unimportant. Both transnational corporations *and* local producers are involved in polluting and risky activities, but it is the transnationals who exercise control.

The specific way in which a transnational controls an operative can be various. Whatever the form, the relationship of control implicates the transnational corporation in more environmental problems than the size or involvement of the company-specific activities would suggest.

In this study we define transnational corporations as the sum of corporately controlled industries and services. We will discuss this in more detail in the following paragraph on "key positions in the production column".

Key positions in the production column

Transnational corporations have subsidiaries or affiliates throughout the world, and have relations with suppliers and buyers. They extract natural resources from one part of the world and export semi-manufactured goods, manufactured goods and services to other parts of the world. They are a very important link in international economic traffic. In their strategic policy, transnational corporations continually assess their position in relation to the "production chains" in which they are active.* The vertical integration strategy is used in some chains, while others opt for horizontal concentrations and still others for acquiring key positions. In all these strategies, the corporate aim is to build and maintain its position of dominance in the chain of production. As a result of this power, transnational corporations play an extremely important role in international trade and, in almost all parts of the world, they are an extremely important party in national and regional economic development processes.

---

* A production chain or production column is a progressive series of steps or phases in the generation of a product. It begins with mining or agriculture and ends with the user. Environmental studies have introduced the term "life cycle", emphasizing that the lifetime of a product is not ended once the consumer buys it, but continues on through to the waste phase, where it is dumped, burned, discharged or recycled.

Economic power and vertical integration in the aluminium industry

Twenty large corporations dominate the world's aluminium, among them Alcoa, Alcan, Reynolds, Maxxam, Pechiney, Norsk Hydro, VIAG/VAW, Alusuisse, RTZ, INI and EFIM. These transnationals owe their tremendous economic power to the vertical integration of their production chain. The corporation is not, however, equally visible in every link of the chain.

Bauxite mining

The large corporations own concessions on "no more than" 30 to 40 per cent of the known world bauxite reserves. In absolute terms, however, the largest concerns, including Alcoa and Alcan, have very sizeable reserves. They can decide for themselves whether, and to what extent, they will mine these concessions. The bauxite reserves of these corporations are found in the countries with the largest bauxite supplies, including Australia, Guinea, Guyana and Brazil. They combine operations in a variety of ways with smaller aluminium companies, forming mining consortia. As a result, the aluminium corporations are able to negotiate very positive mining terms in each of these countries.

Alumina production

The corporations can not only decide for themselves when and where they mine their own bauxite concessions, but also have a large hand in the next link of the global supply chain: the production of alumina. Eighteen corporations own almost 90 per cent of the western world's alumina production. These same eighteen corporations are involved in bauxite mining.

With this scale of alumina factories and concentrated ownership, the corporation knows with every new mine it opens exactly who will buy the bauxite and how much they will buy. Conversely, in opening a new alumina factory or expanding the capacity of an old one, the corporation always knows which bauxite it will be processing.

Aluminium production

A mere 20 corporations control almost three quarters of the world's aluminium production. Most of them are actively involved in bauxite mining and alumina production. From the late seventies and throughout the eighties, more and more smelters were constructed in Third World countries, while investments increased in Australia and Canada. Brazil, Venezuela and Indonesia became the most important countries for new investment, together with the Middle East, Qatar and Bahrain. The price paid for electricity is one of the most important determinants in the choice of where to situate a smelter. Most of the production of these smelters is for export to traditional user areas – the United States, Europe and Japan. The smelting capacity, particularly in these export countries, will continue to grow.

Manufacturing to semi-manufactured and manufactured products

The first step towards manufacturing the final product – casting (alloys), moulding, extrusion and rolling – is almost entirely in the hands of corporations which have their own smelting capacity.

In the second stage of manufacturing, from semi-manufactured product to manufactured product, the aluminium producing corporations are strongly represented only in certain business segments. The corporations are actively involved in a number of existing, highly valuable applications, such as specialized extrusion of profile mouldings, the production of aluminium plate for drinking cans, construction materials etc. The corporations are also actively involved in "new" applications, such as ceramic products, aluminium-synthetic composite materials, and alloys. Market growth in this sector is primarily determined by the technological development of the products. The corporations invest tremendous amounts in research and development, and promoting the applications of the new products.

Because transnational corporations have powerful economic positions in polluting and high-risk production areas within production chains, they are important on two accounts. They are a link between economic development and underdevelopment, and between economic and ecological problems. Transnational corporations embody, as it were, in their worldwide activities, an important part of the relationship between development, underdevelopment and ecology. In other words, between wealth, poverty and environmental problems.

It will be clear that "sustainable development" has everything to do with the doings of international concerns. Without transnational enterprises there will be no sustainable development. That seems unquestionable. The controversy over their role concentrates on the question of whether the corporations are willing and able to take their own responsibility or whether they should be forced into acceptable environmental behaviour. Many aspects of the international activities of the corporations influence "sustainable development". One of the subjects of discussion if the *freedom* of the international concerns *in their choice of location of (environmentally polluting) investments* and the international *geographical shift in the environmental problems* which will result. The following chapter will deal with this in more detail.

Summary

The extent of the involvement, responsibility and liability accredited to transnational corporations depends on the way they are defined, and what one means by being influenced by transnational corporations. These business entities are prominent in environmentally sensitive sectors of the economy, such as mining, chemical production, heavy metals, wood paper, agro-business and the oil industry. A large segment of the production is concentrated within a limited number of corporations. Because transnational corporations are in positions of economic power in production chains which include polluting and high-risk production areas, they form an important link between economic development and underdevelopment and economic and ecological problems even when they do not exercise actual ownership at all stages of production. Thus, they can be held largely responsible for environmental problems in many chains of production.

CHAPTER 4

The Third World as alternative location

Relocation or shifting production

The most important element of our economic system is the freedom entrepreneurs have to trade and invest where and when they want. National governments establish varying limitations to this freedom.

Characteristically, transnational enterprises are able to choose between international locations for their investments. How transnational corporations use this freedom is an important touchstone in judging whether they are able to develop environmentally healthy activities *on their own*, that is, without internationally binding regulations. The economic theory of international comparative cost advantages suggests they cannot. According to this theory, transnational corporations will be most likely to choose to set up production plants in Third World countries that do not insist on strict environmental legislation similar to that in the North, thus offering low-cost (and high pollution) production opportunities in comparison to Northern sites. However in practice this situation is less simple.

First, there are many factors influencing the choice of a certain investment or location. Environmental costs are one factor, as are infrastructure, the geographic position in terms of supply of resources and outlets, the availability of qualified personnel and the structure of the concern (position in the chain of production). Environmental costs usually consume a relatively small part of the company's budget (although this varies according to the sort of company). In practice, the low environmental costs of a potential new plant location seldom turn the scales in chosing to invest in that location.

Liability for land pollution

While the liability laws relating to crimes against the environmental vary from country to country, there is a definite trend in the North towards introducing stricter legislation. Strict liability legislation in relation to surface pollution directly influences the conduct of an enterprise in choosing a location for its activities and investments. In the North, companies always commission an environmental surface study, and they sometimes also require the previous owner to furnish them with a guarantee of non-liability, before deciding to purchase land and buildings. If they do discover that the land is polluted and if they are not guaranteed protection against liability claims, they will not purchase the property.

The fear of being pursued by damage claims is not unfounded. In the Netherlands, for example, there have been many examples of companies causing very serious surface pollution, sometimes many years earlier. This damage totals between 126 and 180 billion guilders, and is spread over more than 100,000 sites. The so-called New Civil Law holds the property owner responsible for all pollution, even if the owner can in no way be blamed (risk liability). The period of limitation of 30 years is not always applicable, and a plea of ignorance not always valid. Hundreds of civil suits are currently in preparation or under review. Those involved face severe financial threats. This has made companies, citizens and the government decidedly more "claim conscious".

In the United States, strict codes of environmental liability have been operational for many years. "Multi-million dollar suits" are not uncommon. In 1988, for example, Shell was forced to pay the sum of one billion dollars in damage claims. Companies in the Netherlands can, since 1985, insure themselves against the costs of environmental liability. A similar "Environmental Liability Insurance" has been available in the US for many years. In the US there is also the so-called "Superfund". This fund provides finances for cleaning up surface pollution in cases where the culprit cannot be found (or where the risks cannot be insured, as is the case for nuclear power plants). The fund is made up of the proceeds from levies, environmental penalties and damage compensation fees.

Transnational companies are not only "claim conscious" in the highly developed countries of the North. Western entrepreneurs hesitate to take over existing East European chemical and metallurgical industries, apparently in anticipation of more stringent environmental legislation. Moreover, industrial equipment in Eastern Europe is so old that bringing it up to date would be too costly.

In the mid-eighties, a "pollution haven" debate surfaced in scientific and political circles. Transnational corporations were accused of "relocating activities" to developing countries with less radical environmental legislation. But prolific research led to the conclusion that environmental legislation was not as important in influencing investment location as economic motives: cheap production conditions including labour costs. An exception to the rule were the producers of highly toxic substances (asbestos, dyes, pesticides) and heavy metals (copper, zinc, lead).[12]

Northern scientists and politicians were pleased with this confirmation of their argument that stricter environmental legislation does not lead to job and industry loss in the North. They reaffirmed that it was possible to conserve both welfare and the environment!

But the matter is not quite so simple. While few companies have actually moved, (i.e. the same enterprise closing a factory in one area and investing in another in another area), production is being "shifted out". When one company decides to stop the (environmentally polluting) production of certain goods, buyers are then forced to look elsewhere. This provides new companies with new expansion possibilities, possibly in the Third World. The end result is that production is "shifted" from one country to another. It is probably more true to say that the world trade is the solution for domestic environmental problems. This process of shifting of production is happening in a number of basic metals industries, in the heavy – bulk – chemicals industry and in other industrial branches. Wherever these production industries move, recipient countries experience rapid industrial growth – and extra environmental problems. Because these branches of industry are growing *faster* in the South than they do in the North, the strain on natural resources there is growing fast, the local environment suffers increasingly, and the public is exposed to more and greater health risks.

Rapid industrial growth in South East Asia

Asian economies (excluding Japan) are growing faster than those of Europe and the United States. The textile, electronics and car industries, in particular, are expanding in size. These types of businesses are important customers for the chemical industry – and the chemical industry (for example plastics and other semi-manufactured products) is flourishing. In 1993 the "chemical market" expanded in Asia (excluding Japan) by no less than 8 per cent, while elsewhere in the industrialized world, growth stagnated. Participants in this growth are both Asian enterprises and transnationals from the United States, Europe and Japan, who invest huge sums in new production units. The Asian market is not the only outlet; through the combination of low costs and modern technology it is possible to compete on the international market despite long transportation distances. This was clearly the case with organic dyes: between 1982 and 1991 the Asian share of world exports of organic dyes rose from 13 to 24 per cent. The tremendous growth of the chemical industry and the shift of the centre of gravity to Asia is illustrated in the following prognosis:

World market for chemical products in billion dollars, and in percentages.

	Bill. dollars		Growth	Percentage distribution	
	1992	2000	%	1992	2000
North America	319	400	25	29	29
Western Europe	430	500	16	40	36
Asia	335	490	46	31	35

European transnationals find Asia attractive because of its growth potential, but there also so-called "push factors". Manfred Schneider, president director of the chemical giant Bayer, described it in this way:

"The main disadvantages we face (in Europe) are higher labour costs and expensive social security systems, coupled with the wide-spread regulation of environmental affairs by the state. We have to overcome these handicaps. If we do not, many areas of our business (in Europe) will become uncompetitive and therefore in danger of being squeezed out of the market."[13]

To conclude that the geographical shift of polluting production processes is not as bad as it seems is not only short sighted, but complacent. It assumes that environmental measures in the North are effective. Unfortunately we all know that the "stricter" environmental policy in the North has not been able to deal with either international or local environmental problems.

It would be more logical to conclude that the badly needed tightening of environmental measures in the North will indeed lead to the exodus of large, environmentally polluting production processes to the South. The example below shows how that possibility became reality in the phosphoric acid industry.

The transfer of phosphoric acid production: closures in Europe and expansion in Morocco

Phosphate, in the form P_2O_5, is a nutrient for plants and a building block in food production. Modern intensive agriculture boosts natural phosphate levels in the soil through the addition of the phosphate fertilizers. The production of this chemical fertilizer is one of the many stages in the food production chain in which this nutrient continually changes shape. Many environmental problems arise throughout the entire chain, one of which is the production of phosphoric acid, which is particularly well-known for the discharge of waste gypsum into surface water and the release of the highly polluting substances, phosphorous, cadmium and radon^{226}, into the environment. For every tonne of phosphoric acid (P_2O_5) produced, five tonnes of unclean phospho-gypsum is discharged.

In the 1980s, Western European governments exerted their authority to stop such discharges, demanding a reduction in the quantities of free phosphorous drained because of its eutrophic interaction with surface water. Many local governments demanded a total halt (mostly in phases) to the discharge of cadmium – an EC blacklisted substance.

At first, the Western European chemical fertilizer industry responded with delay tactics. They began researching alternative methods for processing the gypsum. They looked into storage on land, as is done in the United States, and into reprocessing the waste into building material. The largest chemical fertilizer producer in the world, Norsk Hydro, initiated an ostensibly serious effort to develop a new cadmium cleansing technique at their plant in Vlaardingen, the Netherlands. The second largest, Kemira, offered to modernize its entire plant in Rotterdam in exchange for a promise by the local authorities to freeze their stricter discharge standards for at least 20 years. In anticipation of research results and investments in cleaner techniques, companies everywhere requested postponement of the application of the stricter discharge standards. However, local governments took measures to progressively tighten rather than freeze the discharge standards.

At the end of the 80's, in order to meet the stricter discharge standards, many producers in Western Europe switched to a cleaner raw material, low-cadmium phosphate rock. In three European locations producers switched to less environmentally harmful nitrophosphate production methods. These were Norsk Hydro in Porsgrunn (Norway), BASF in Antwerp and Chemie Linz (Austria).

The Western European phosphoric acid producers were also riddled with economic problems. In 1988/89 the Moroccan state-owned concern, OCP, started expanding its production capacity since Morocco can benefit from cheap phosphate ore mined on its own territory. A global glut arose at the same time that Western European demand for phosphate fertilizer shrank and prices begun to dive. Cheap imports from North Africa snatched the market from European producers and Morocco is still expanding capacity to meet export markets. In 1996, total world phosphoric acid production capacity will have risen by 1.9 million tonnes.

Along with the continuing decline in the use of chemical fertilizers in Western Europe, this bodes ill for the European phosphate fertilizer industry. These negative market prognoses have forced corporations to close production units. From January 1992 to April 1993 alone, 1.35 million tonnes of P_2O_5 phosphoric acid production was abandoned in Western Europe. That is more than one third of the total capacity, a decrease far in excess of the decline in the market and the oversupply caused by the expansion of competing nitrophosphate fertilizer production. Imports of phosphoric acid (or phosphate fertilizer) from Morocco (in particular) will have to be brought in to meet the needs.

But the result will be a surge in pollution in Morocco. The flood of highly polluted phosphoric gypsum is being pumped, untreated, straight into the Atlantic Ocean. Thus closures in Western Europe have only resulted in a shift of pollution to Morocco. In the end the environmental cause has gained nothing.

This illustrates the dubiousness of transnational corporations accepting "environmental responsibility" and choosing not to move when confronted by tighter environmental legislation in the North. Tighter environmental measures there, combined with the continuing absence of international legislation, only makes the South more attractive. These same transnationals claim only to be able to take on their environmental responsibility in the continued absence of international legislation.

We may therefore conclude that while the differences in environmental policy between Northern and Southern countries is seldom a motive for transferring production, it is the lack of environmental motivation in the closing down of company activities that indicates the inadequacy of the environmental policy. If governments in the North will significantly tighten environmental regulations, which truly attempt to solve environmental problems, this will result in the transference of polluting activities to the South. This underlines the need for international environmental legislation.

The great negotiating power of transnational corporations

The above conclusion assumes that the differences between the North and the South in environmental legislation and in the enforcement of the rules and regulations will remain, at least for the time being. In many cases, this is not only due to differences in formal legislation or in their enforcement, but is related to the differences in the negotiating power of the pertinent Northern and Southern government agencies involved. At a regional and local level, specific environmental regulations and their application are largely the result of negotiations between companies and government agencies. Transnational corporations can have an immense influence on the results of these negotiations due to their freedom to choose between various investment locations, specific knowledge and their financial resources.

Countries in the South are largely dependent on foreign capital for investment in economic enterprises. In the case of large infrastructural projects or major direct investments in mining or processing, transnational companies are always involved. This is certainly the case for huge projects which go beyond the financial capacity of the country, and is usually demanded by international lending agencies such as the World Bank, who never fully finance development projects but require additional investments by the actual operating company as well as their technical and organizational expertise. It is almost inevitably a transnational corporation that can meet these financial, technical and organizational demands. Large investment programmes in mining or basic industry is therefore always a complex structure consisting of national governments, international banks and transnational corporations, who negotiate their contributions based on their individual interests in the project. The World Bank usually coordinates the search for finance, the Third World country is usually the project applicant in search of money and expertise to stimulate economic development, while the transnational is the potential investor in search of profit. If the Third World country has a lot to offer, for example an abundance of natural resources and favourable government subsidies, an attractive cooperation agreement and cheap energy supplies, this will improve its bargaining position and it will be able to make far-reaching demands, for example concerning local content in production, local employment and skills transfer and so forth.

Footloose industry

If a Third World country with few natural resources wants to attract foreign investment to improve employment opportunities or stimulate export, it will only succeed if the government is flexible, creating incentives such as free trade zones, favourable tax regimes for foreign business and a relaxation of national labour laws that affect, for instance, trade union membership or conditions for women workers. These attract mainly smaller investors who are not interested in specific natural resources but are involved in labour intensive production and are therefore particularly concerned about labour costs and social stability. The transnational corporation is almost always in a position to invest elsewhere, unlike national companies which are limited by national boundaries, so it can make demands over taxation regimes (exemption from or the application of low tariff taxes, special import and export duties), repatriation of profits, mining conditions (applicable to mining projects: concessions and royalties), and the presence of an infrastructure which is financed by the host country.

Energy politics of aluminium companies in the Third World

In the past thirty years, several large hydro-electricity projects have been established in Third World countries, for example in Surinam, Ghana, Indonesia, Venezuela and more recently Brazil. The countries involved aimed to use this type of electricity production to stimulate long-term economic development. Because of the massive investment required for by the construction of a dam, a reservoir and power station, this sort of project is only viable if its entire output is purchased from the very start, an impossible condition for most developing countries where per capita electricity consumption is usually very low. The sorts of business activities that are stimulated by a hydro-electric plant are not present and can only be built up slowly if electricity tariffs are not too high. But a modern aluminium smelter with a capacity of 200,000 tonnes per year easily consumes 310 Megawatts of electricity and is therefore a perfect consumer. In most cases, an aluminium industry therefore becomes a precondition for launching electricity production projects in the Third World. The two go together like Siamese twins.

The large aluminium corporations often negotiate the establishment of *one* aluminium smelter in various countries at the same time, thus creating a powerful negotiating position for themselves. In many cases they are able to force the electricity tariffs down to – or under – the cost price by pitting countries against each other.

Often, however, the disadvantages of combined hydro/aluminium projects are greater than the advantages. The income from export and the increased employment, which undoubtedly have repercussions on the economy, have to be weighed against the destruction of the natural environment caused by flooding huge areas of land, salinification in downstream areas, reduction in soil fertility due to decreased silt deposits, more water-related illnesses (such as bilharzia and malaria) and a reduction in fishing harvests. The indigenous population is usually forced to re-settle with little or no compensation. Moreover, this sort of project usually leads to debts lasting 30 to 50 years and capital goods and services (repairs/maintenance) usually have to be imported. Often, corporations are able to profit from tax perks and, in many instances, they are free to repatriate their profits.

Numerous investment projects have shown that the negotiation structure involving government, international banks and transnational corporations takes insufficient account of the interests of local population groups and the natural environment. From the economic point of view, the value of large scale projects for economic development has been the subject of criticism for many years. In the 80's environmental groups and interested groups of local and indigenous populations added their criticisms.

This has focused particularly at the World Bank as external financier.* It is also directed at the governments concerned. But

* As a result of which the World Bank altered its project review procedures. An examination of projected environmental consequences of the investment is now standard procedure.

interest in the environment and associated problems of poverty usually have to give way to economic interests.* Only in unusual situations, often when there is an environmental scandal, does the role of the transnational come under fire. The powerful position of the international companies in tripartite negotiations with banks and governments is almost never the subject of discussion.

Summary

One of the most important freedoms of transnational corporations is their freedom to choose the location of their investments. It seems likely that they will not be able to keep to their self-declared environmental responsibility and will avoid the tighter Northern environmental legislation by relocating their production facilities to the less strict Third World. Research indicates, however, that only under exceptional circumstances will corporations relocate their production activities to countries with less stringent environmental laws. To conclude, though, that the migration of polluting processes of production is less of a problem than was anticipated is both short sighted and conceited. That conclusion rests on the supposition that environmental measures in the North are effective. Unfortunately, we know only too well that the "tighter" environmental policy in the North has not been able to conquer either the international or the many local environmental problems. It would be more to the point to conclude that the necessary tightening of environmental measures in the North will lead to an exodus of polluting production processes to the South. This process can influence the welfare of the North while it will mean a large increase in the environmental problems of the South. Through their freedom to locate plants and investments, transnationals will be constantly in the position to avoid their environmental responsibility.

---

* The interests of the local population are often parallel to those of the environment.

While systematic relocation of polluting production processes to "pollution havens" does not occur, there is a relative transference or shift. We see an increase in the share of the South in various industry branches – including branches which are known for their large scale environmental pollution. A faster growth of these industries in the South than in the North is combined with an increase in the demands placed on natural resources, with large negative consequences and risks for the local environment and public health. Transnational corporations are also involved in rapid expansion in some of the countries in the South. Through their tremendous negotiating power, they are usually able to demand extremely favourable location and investment conditions. Almost without exception, they are unwilling to apply the stricter environmental laws of the parent country to their location in the South, where laws are less demanding or law-enforcement less effective. In other words transnational corporations adapt themselves to the lowest pollution standards in their investment behaviour. They do not create an example for the surrounding business culture. There is presently no international legislation to correct these tendencies.

CHAPTER 5

Environmental technology as solution

Foundation for progress

Technological innovation is an economic necessity. It enhances the productivity of companies, which in turn contributes to lowering the cost price. It also enhances the quality of a product. If a company wants to compete on the basis of price and quality, technological innovation is a prerequisite. The industrialized world strongly believes in the blessings of technology, seeing it as the vanguard of economic progress and welfare, and, currently, the answer to environmental problems.

Transnational corporations own ninety per cent of all technology and product patents in the world. (Only six per cent of the 3.5 million patents in the world are owned by transnational companies in developing countries[14]). It is little wonder then, that they have such an important role in developing, applying and disseminating environmentally friendly technology in the South.*

For analytical purposes, it is important to distinguish between production-integrated technology and "end of the pipe", or purification technology. Environmental technology is production-integrated if it is an integral part of the production process; when it is only applied externally to the production process, to modify emissions and waste discharges, it is known as purification

* The term technology is a general term. We use it here to mean both knowledge and techniques, which aim at preventing and reducing environmental pollution and its risk.

technology. This latter is always added to the process and therefore always increases the cost price of the product, whereas production-integrated technology is preventative and may lower the cost price. Because it raises costs, purification technology is almost always applied to meet government requirements concerning pollution.

The question we have to ask, however, is whether it is wise to leave progress in the area of environmental technology to the voluntary initiatives of transnational corporations.

The fundamental limitations of company technology

From looking at a finished product you cannot see how environmentally sensitive its production process has been. Consumers are not willing to pay extra for a cleaner product if they can't see the damage a "dirtier" product causes to the environment (and to the health of the workers). They give preference to a cheaper product from a competing firm, even if it was produced under worse labour conditions and in an environmentally damaging way. Good behaviour is not rewarded. Bad behaviour is. The result is that producers and consumers shift as many of the costs as possible onto society and nature. Even when a product visibly pollutes when used, the average consumer is not willing to pay extra for a less polluting product. Thus, the market mechanism, working from these incomplete signals, will always act as a constraint to sustainable development.

This same defective market mechanism also determines which form of technological renewal is important to the business community. Corporations will only introduce new technology if it promises lower costs and increased sales. This applies to all technology, including environmentally friendly technology and, in particular, to *production-integrated* technology. This economic principle not only determines the application, but also the development of new technology; any technology developed on other principles will simply not sell in the long run. The technology that finds its way to the work floor or the market is not necessarily the best (environmental) technology, it is the one that will not increase but rather decrease the price of a product.

The marriage of financial and environmental advantages

The post World War Two history of pot oven technology for aluminium smelters provides a perfect example of the marriage of financial and environmental advantages in the development of technologies.

A modern aluminium smelter is capable of producing approximately 200,000 tonnes per year and requires 300 megawatts of electric power. The average western, coal-fired modern electric power plant has a capacity of 600 megawatts. An aluminium smelter which uses the modern 175,000 ampere pot oven technology developed by Pechiney requires approximately 13.5 kilowatt per hour per kilogram of aluminium. Previously, pot ovens that were used (1940 to 1955) required approximately 19.4 kilowatt per hour per kilogram of aluminium. This improvement in energy-efficiency has reduced the cost price of aluminium, while other improvements in pot oven design have decreased costs of construction.

They have also resulted in considerable reductions in fluor and fluoride emissions, which were 15-25 grams per kilo of aluminium produced in 1940–1955 and are now 0.5-1 gram. By introducing purification technology, sulphur dioxide emissions were also reduced. Only CO_2 emissions remain high. But none of these changes would have been made if they didn't reduce the cost price.

The inherent economic limitations to production-integrated environmental technology, developed and applied by the industrial community, ensure that it will not provide solutions leading to pollution prevention or reduction. The aluminium smelter example demonstrates that business economics do provide openings for progress from an environmental point of view, but these improvements have not, as yet, been able to reduce CO_2 emissions, and fail to solve global and many national and local environmental problems.

Fundamental and far-reaching intervention is required for solving present day environmental problems. It is unrealistic to assume that the business community will voluntarily develop the necessary environment-technical solutions.

Unattainable environmental technology

This is not the only problem. Technological know-how is unevenly distributed throughout the world: the newest technology – and particularly production-integrated environmental technology – is either inadequate or financially unattainable in developing countries.

There are three major phases in the development and application of production technology:

- The transnationals design and develop the new technology in specialized Research and Development units, generally located in the North, where the highly educated personnel needed for this work live. Some of the basic scientific research and laboratory experiments are also conducted by universities and government-subsidized organizations in the North.
- The new technology is usually scaled-up in one of the company's existing plants close to home, i.e. in a developed country, and is largely adapted to the scale of production, environmental conditions and market size there, as well as to patterns of consumption and consumer demand.
- Later, the technology is distributed throughout the world. One of the most important goals is to acquire a return on investment on R&D and production costs within the shortest possible time. If no competitive technology exists, there is a strong tendency to patent the technology, to protect it, and disseminate it only within the transnational which owns patent. If there are competitive technologies available, the company will license the technology to ensure maximum return on investments.

Licences

A licence is the right to use a specific technology. Licenses are widely used in the world of chemical production, where corporations sell, not only their knowledge of production in the form of licences, but also calculate the cost of the licence into their own costs of production. The general practice is that the company using that particular technology pays the owner of the technology, a sister company, for the licence. In this way licences do not interfere with competition or inhibit technology transfer, and provide a flow of income into the company that owns the licence.

This process of developing, introducing and spreading technology has two important consequences. First, per definition, existing production installations (especially in the South) are always outdated, due to the continual development and introduction of new technology in the North. Older and outdated production installations cause more pollution than can be justified by the state of technology. Secondly, technology cannot simply be transplanted, unchanged, to the South – to countries with other socio-economic, ecological and climatological circumstances.

These two points are frequently brought up in the debate on the transfer of (environmental) technology to the South. We will look at them in more detail.

Double standards in environmental technology?

A limited, but growing, number of transnationals apply the same environmental standards to their global activities. Recent management literature indicates a trend in the direction of globalizing company environmental management, and of companies basing their world-wide environmental standards on the strictest laws. The most important impetus is the need to minimize liability and avoid a confusion of environmental standards within one corporation, rather than an increased corporate environmental awareness.

Agenda 21 recommendations

There are several references in Agenda 21 to the need to create identical international environmental standards within transnational enterprises, for example, Article 30-22 of the chapter on Business and Industry:

"Business and industry, including transnational corporations, should be encouraged to establish world-wide corporate policies on sustainable development, arrange for environmentally sound technologies to be available to affiliates owned substantially by their parent company in developing countries without extra charges, encourage overseas affiliates to modify procedures in order to reflect local ecological conditions and share experiences with local authorities, national Governments, and international organizations."

The Benchmark Survey carried out in 1991 by the United Nations also shows that while only a handful of transnationals intend to introduce this sort of policy, still fewer have actually done so.[15]

The concept of global environmental management is a step in the direction of sustainable development. Put differently: if the transnationals apply the same environmental technology and practices in Third World countries as they do in the parent country, this is often a step forward relative to the situation in which they only heeded local government laws.

But it is no more than a step. Business economics, we noted earlier, defines what the "best available environmental production technology" for global application is. This is not the same as the best solution in terms of the environment. In practice, it also means that a company will only introduce and apply better (production-integrated environmental) technology in its *new* investments throughout the world. Existing installations will avoid having to apply the new environmental technology.

No single production process is clean. Purification technology is an environmental necessity. Because it increases the cost price of the product, companies tend to adapt their purification technology to the demands placed of the environmental laws on emission and discharge standards in the host country. This places high demands on the efficiency of regulatory bodies. Local standards and the way they are enforced differs from country to country. These circumstances make it all too easy for transnationals to delay introducing a general policy of applying the best purification technology throughout all their plants.

Unadjusted environmental technology

The application of the best production-integrated environmental technology will not necessarily lead to the least pollution, for various reasons.

- The breakdown-sensitivity of the technology. Break-downs are the cause of most emissions and risks. In chemical production industries, for example, there are variable conditions from one country to the next. Technology can be affected by climatological differences and the quality of raw material, cooling water and electricity supply. Advanced forms of preventative maintenance that are common in the North are often not available in the South.
- The speed with which disorders are repaired also varies from region to region. Northern technical teams often have to be brought in to carry out specialized repairs in the South. This is not only expensive, but time-consuming. There is often a lack of spare parts (due to the dearth – or absence – of other industries and dependence on distant suppliers).
- Inadequate grasp on the technology. In general, the South has fewer specialized, highly educated personnel at its disposal and there is often an inadequate grasp of the technology.
- Geography also plays a part, through the differences in vulnerability of the natural environment. The toxic activity of complex chemical substances, for example, varies under different climatological conditions. The effectiveness of agricultural herbicides has a shorter duration in the tropics than in cooler climates. In the case of dam construction, however, the tropical

natural environment is more susceptible than a temperate one. Artificial lakes for hydro projects in tropical and subtropical regions house various insects and disease-bearing organisms which cause public health problems to local populations. In warm, wet regions, toxic wastes often seep through waste storage deposits (this often occurs with phosphate gypsum stored on land). The so-called pond or sedimentation system (used in depositing various mine tailings and the toxic red mud in alumina production) is more vulnerable in the warm, wet tropics than elsewhere.

Ecological carrying capacity and international environmental standards for companies

There are regional differences in the vulnerability of ecological systems. The term "ecological carrying capacity" denotes the varying levels of pollution which different areas can tolerate before the environment's natural regenerative ability is affected. This suggests that it is possible to vary pollution standards from region to region and that an internationally operating company can justifiably apply different environmental standards in different regions. However, it is an unsound conclusion in the context of economic development and the scientific concept of "eco-space". If *one single* company is allowed to pollute to the extent of the ecological level of tolerance, that company lays claim to all of the locally available ecological usage area. In other words, no other industries can come to the area, and there is no room for economic growth. In terms of "sustainable development" (which aims at integrating both ecological and economic ideas), it is preferable to have globally uniform environmental standards based on the best technical solutions.

We can conclude that, as the environmental risks and repercussions of a particular technology differ according to region, new and already known environmentally sound technological solutions have to be adapted to the local situation in order to guarantee the best results.

Redundant technology

Industrial companies are links in global chains of production which begin with the extraction of raw materials and ends with the consumer, and, in the most positive scenario, the recycling of wastes and their return to the chain. The manufacturing process brings emissions and environmental risks at each stage of the chain. In separation, refinery and extraction processes, a cleaner raw material produces fewer toxic emissions (and risk of emissions). Similarly, if the more chemical and hazardous substances are extracted from the raw material at the beginning of the chain, there will be fewer environmental risks and effects at the end of the chain. (In the case of synthetic production, the purity of raw materials is a less important factor.) As a rule, the sooner environmental technology is built into the production processes of the manufacturing chain, the more effective and more preventative it will be.

Cadmium purification at the source

Phosphate rock, mined in Togo and Senegal, has the highest cadmium-holding percentage in the world (172-234 mg of cadmium per kilo). Togo exports all the phosphate it extracts, and 20 percent of its government income comes from this source. In Senegal, some of the phosphate ore is processed into phosphoric acid and phosphate fertilizer.

From 1986, phosphate export to Germany, Belgium, the Netherlands and other European countries declined, after the introduction of tighter European and national environmental laws. These countries gradually lowered the permitted levels of cadmium in production wastes and produced goods. Fertilizer and feed-phosphate producers switched to the "cleaner phosphate ores".

The governments of Togo and Senegal applied for an EC grant (from the Sysmin fund) to research the purification of phosphate ore. Without success: at present, while it is technically possible to remove the cadmium, it was not economically viable. A Danish engineering company reviewed the technology.[16] At a pilot plant in Taba, Senegal, they were able to remove 80 per cent of the cadmium from the . The Togolese government failed to interest foreign investors in a 500-million-dollar phosphoric acid/phosphate fertilizer plant.

As we can see from the above example, much of the end-of-chain purification technology becomes redundant if purification occurs earlier. Corporations are not necessarily interested in this principle. No government regulations require companies to use pure or purified raw materials. Most government regulations merely aim to limit damage. There is no environmental policy based on "integral chain management" and companies are completely free to choose where they buy their raw materials. Pure raw materials are usually more expensive than impure ones. Thus, only large, vertically-integrated transnational corporations, or those with a key position in the production chain are able to shift the application of environmental technology to the first stages in the production chain.

Inadequate technology

Applied environmental technology can have tremendous results. As we have seen, fluor emissions from aluminium smelters were drastically reduced over several decades and productivity increased considerably. These sorts of changes in the level of pollution have been observed on the local level in the industrialized North over the past ten years. However, the number of production locations has expanded globally, bringing new sources of local pollution to more areas. Wherever industries are established, they are accompanied by the destruction and pollution of the local environment. Because of this process of uncontrolled production growth, regional and international environmental problems are actually increasing instead of decreasing. The growth in production in new locations not only effects local air and water but increases global problems of environmental destruction, exhaustion and the problem of waste.

To embark on the road to sustainable development, therefore, the expansion in production has to be accompanied by an absolute reduction in environmental pollution in *the entire production chain.* Very few large concerns have set themselves the task of reducing their *absolute* levels of emission and risk. The majority continue to measure the success of their environmental policy by the *relative* efficiency of applied purification technology.

Summary

Transnational corporations have a central role to play in the development and dissemination of environmental technology. The inherent limitations and deficiencies of their perspectives lead us to conclude that sustainable (industrial) development cannot be achieved if the corporations are entirely free in developing initiatives:

- transnational corporations develop and apply only profitable (environmental) technologies. In corporate strategy, contributing to sustainable development will always be of secondary importance.
- the development and introduction of new technology mainly takes place in the North and conforms to the scale, potential and surroundings of production in the North as well as with the size of the market, quality demands of buyers, and consumer behaviour there. New (environmental) technology will only be disseminated, under licence, if there is a competing technology. Otherwise the transnational will protect the technology with patents.
- transnational corporations rarely apply the best technology throughout all their divisions. Investments in the newest and best technology only occur when new installations are constructed.
- the application and effectiveness of purification technology depends largely on the demands of local governments and on the level of enforcement of local legislation.
- the same technology will have different risks and consequences for the environment in different regions. This means that only local adapted – technological – solutions will lead to the best results. Modern environmental technology is, however, mainly applied without proper adaptation to the local situations.

- only when expansion in production is accompanied by a reduction in environmental pollution throughout the entire chain of production will we have started on the road to sustainable development. Although many transnationals hold strategic positions within production chains, this is not expressed in preventative management through the application of as much environmental technology as is available at the earliest stages of the chain.
- current environmental technology is often inadequate because the successes in reducing pollution are outweighed by the increases pollution caused by the expansion of the number of polluting industries in the world. Very few large corporations have – and only recently – set themselves the goal of decreasing emissions and risks. The majority measure the success of their environmental policy on the relative efficiency of applied purification technology.

CHAPTER 6

Are environmental policies focused on sustainable development?*

If the initiative for sustainable development is to be left to the transnationals, then their environmental management policies will have to reflect their alleged good intentions. These policies should be so strong that they transform company activities in the direction of sustainable development.

Company environmental policy is changing fast. In recent years, transnationals have become increasingly aware of the negative environmental effects of their activities. There are two main reasons for this:

- Their awareness of the increasing intensity and size of the environmental problem itself.
- An increasing governmental and non-governmental pressure to regulate and control polluters.

The central question in this chapter is whether corporate environmental management really aims to create sustainable development. There is no question that through their international activities, transnationals can play a crucial role in this direction. Agenda 21 demands:

> "The business community, including transnational corporations, should recognize that environmental management is one of the highest priorities and a decisive factor in sustainable development."

---

* We use the term environmental policy as synonymous with environmental management. In our view, policy is a form of management. We avoid using such terms as "environmental administration" or "care for the environment", which mistakenly imply that corporate environmental policies care for or administer to the environment.

Can and will the transnational corporations come through on this?

The United Nations published a study in 1993 entitled "Environmental Management in Transnational Corporations."[18] This report, better known as the Benchmark Survey, brings to light the present state of environmental management in 210 transnational corporations with an annual turnover of more than 1 million dollars per year.* Analysis of the questionnaire responses identified four levels of corporate environmental management based on compliance with environmental interests.

- Compliance-oriented management (the reactive corporation)
- Preventive environmental management (the lean and precautionary corporation)
- Strategic environmental management (the opportunity-seeking corporation)
- Sustainable development management (the responsive corporation).

For each sort of corporate environmental management there are corresponding company activities (environmental measures) and specific governmental measures (see the following table). We will not go into the relationship between the two.**

---

* The research was carried out in 1991. Almost 800 corporations were approached. While this study is the most detailed of its sort, the low response may indicate that it does not reflect the average situation, but more likely the best that transnationals achieve in the area of environmental management.

** The authors report an inseparable link between company activities and governmental policy and regulations. They assume that, in order to introduce more far-reaching forms of environmental management within the entire transnational corporation, matching governmental management must be established, and that a legal framework, tough monitoring and a stable government policy in the corporation's 'home' country are the most important inducements to voluntary initiatives for environmental management. (Benchmark Survey, page 169). Chapter 7 discusses international environmental laws and regulations.

The four levels of corporate environmental management

Management type	Corporate activities	Supporting government activities
1. Compliance-oriented management (The reactive corporation)	– end-of-pipe solutions – abatement procedures – monitoring – compliance reports – training – emergency response	– command and control – realistic regulations – involvement of business in designing regulations – inform on regulations – tough enforcement
2. Preventive management (The lean and precautionary corporation)	– internal audits – pollution prevention – waste minimization – public information – energy conservation – green accounting	– increased liabilities – waste treatment requirements – restrictive landfills policy – community right to know – energy conservation – taxation
3. Strategic environmental management (The opportunity-seeking corporation)	– public dialogue – external audits – disclosure – environmental management integrated in the planning – cradle-to-grave policy – Green R&D – setting environmental health and safety targets	– stable regulatory build-up – green labelling programmes – support of consumer and green investor programmes – market means of regulation – voluntary regulations – R&D tax-breaks
4. Sustainable development management (The responsive corporation)	– developing country programmes – ethical sales policies – international disclosure – climatic change policies – afforestation programmes – world-wide policies – international auditing	– international information dissemination – integration of sustainable development objectives in decision-making – international harmonization of environmental regulations and/or standards – international taxation

Compliance-oriented management

Over the past twenty to thirty years, governments in most industrialized countries have been developing environmental policies. Industries have slowly adapted their practices to the changes, and have often strayed from the letter of the law, assisted by legal ambiguities and weak enforcement. On introducing new environmental laws, government agencies almost always issued permits to continue releasing effluents and emissions to the air at the old level, effectively legalizing these practices. When they applied for emission permits for new production installations, transnational corporations were able to negotiate, locally and regionally, the sort of environmental technology that was to be used and drainage standards to be applied. They were given a lot of leeway. While the body of environmental rules and regulations continued to grow, companies managed to use their influence against rapid application of tougher local and regional pollution standards.

Shell Pernis: promising a cleaner future

In the seventies and eighties, large chemical concerns were able to undermine the tighter emission permits by linking them to new research into environmental and cleaner production technology which progressed particularly slowly causing time limits to be constantly extended and delaying the deadline for the application of standards.

A 1989 SOMO study of the environmental practices of Royal Dutch Shell Chemicals in Pernis, Holland, concluded:

"Using the delay tactic of conducting research into improving hazardous waste discharges, Royal Dutch Shell Chemicals was able to continue dumping unlimited quantities of organic chloride substances and maximum amounts of drins and fine ash into the surface water. It was difficult for people in the neighbourhood and environmental pressure groups to get a grasp on and have a say in the procedures surrounding the issuance of a new discharge permit for extra environmentally toxic substances into the surface water (these substances were not included in the 1987 permit). Shell-Pernis claimed all the time and space it needed to decide whether and when it would convert to reducing or stopping emissions of environmentally toxic substances into the surface water. It has also given itself as much time as it needs to determine the speed at which it will stops drins, organic chloride substances and light ash emissions."[19]

In developing countries, it is usually difficult to develop, introduce and enforce an entire set of standards and rules. The relative scarcity of scientific personnel and financial resources often leads to dependence on the expertise of the transnational corporations themselves, certainly in the area of technical training. This makes it quite difficult to independently enforce legislation.

Environmental rules and regulations in the industrialized North have steadily improved over the past decades. In a previous chapter we saw that companies have scored striking successes in reducing pollution, at the local level. The above example, however, illustrates that environmental management solely aimed at conforming to the rules and regulations has little value. The company was able to undermine the emission permits for new productions units. Its true purpose was to create as much room for manoeuvre as it could.

It is, however, vitally important for transnational corporations to explicitly develop company-wide policies to comply with national and local environmental rules and regulations. All the more so for their operations in developing countries, where governments don't always have the resources to monitor and enforce their own national laws. Unfortunately, few transnational corporations actually have such a company-wide policy. The authors of the Benchmark Survey noted a "disappointing number of examples of transnational corporations which explicitly referred to international conformity in their environmental aims, although more than half of the companies researched have activities in developing countries."[20] The claim that transnational corporations will voluntarily develop environmental initiatives, to adhere to international environmental rules and regulations, can be treated, it would seem, with a grain of salt.

The Benchmark Survey observed that corporate environmental management used six policy instruments in complying with regulations: end-of-the-pipe solutions, reduction measures, the monitoring of emissions, issuing reports on levels of conformity, training and the development of disaster plans. Most corporations in the Survey applied at least one of these measures, which are aimed solely at controlling and managing the releases of polluting substances to conform to standards and regulations, within a given level of pollution in the production technology and a given organization of the labour process. They are all measures taken once the damage has been done to confine the environmental damage to the legally prescribed proportions.

Disaster plans and risk analysis: The European Seveso Guideline
The disasters at Union Carbide in Bhopal, at Seveso and at Sandoz in Basel have had an enormous influence on the safety policy of corporations as well as corporate an governmental disaster plans and projections in Europe. If they had not already done so, most concerns appointed general safety coordinators to coordinate factory safety, the flow of information to the local population and disaster planning. These personnel coordinated safety activities world-wide, and their contact with the corporate head was consolidated. Many companies improved and intensified safety, health and environmental auditing, and responsibilities were defined more carefully. More alarm and detection systems were installed. These measures strengthened the involvement of the head office in the coordination of safety problems, in particular in the higher risk branches of the company, and heralded a shift in emphasis towards anticipation and prevention.

The 1985 Seveso Guideline, which prescribes disaster planning and risk analyses for European companies, is a clear example of mandatory environmental measures at the European level.

Preventive environmental management

Compliance-oriented environmental management is a direct reaction to legislation; preventive corporate environmental management goes one step further. This system relates environmental management to cost-savings, the financial risks inherent in pollution, and the financial advantages of preventative environmental management. "Pollution prevention pays" is the appropriate slogan used for this type of corporate environmental policy.

The economic opportunities and the potential and financial risks of pollution encourages most large companies to introduce some form of preventative environmental management. The increasing risk of being held liable for damages (particularly in soil pollution and environmental disasters) and the possibility to reduce costs (energy savings and waste restrictions) have been the greatest impetus for this management system. Not surprisingly, transnationals with the highest liability risks – the chemical industries – are the most intensively involved in preventative environmental management. The Benchmark Survey found that almost 70 per cent of the corporations in their study had developed programmes for waste restriction, energy savings and accident prevention. These programmes include internal audits, which aim to avert disasters through signalling high-risk and dangerous situations. In these situations, environmental management usually works in close cooperation with safety and health management at the work place. In recent years, modern electronic information technology has provided many new opportunities for monitoring and quantifying environmental effects. "Green accounting" became available to corporations. This is the weighing of environmental effects against the costs of environmental management. This link to financial accounting does, however, have its drawbacks. When a company activity is contracted out to another firm, this activity is no longer the responsibility of the first company and no longer part of the process of weighing green priorities. If a company decides to contract out transportation, it will need less energy (diesel fuel). If it contracts an administration firm for salary administration, it will use less paper. Green accounting does not show that this is simply a diversion of activities from one company to another and that no gain has been made in terms of the environment.

The Benchmark Survey found that one third of the companies interviewed use some form of green accounting.

United States companies in particular tend to apply preventative environmental management, using internal audits, risk analysis and related preventative activities, since when they are found liable for damage, the penalties are high. The Superfund legislation has been a particular impetus to change company policy on environment, safety and health.

The United States Superfund

The Superfund was introduced in the United States at the beginning of the seventies to finance surface clean-ups for damage done where no culprit can be found or when the risks cannot be insured, is the case with nuclear power plants. The fund is fed by levies, the yield of environmental penalties and damage compensation.

In the spring of 1994 the Clinton administration opened negotiations with representatives of employers and environmental organizations to change the Superfund law. The intention is to alter the following parts:

- A pro-rata liability. Instead of holding the corporation that owns the land totally liable for site clean-up costs, the company will be fined according to the extent of the pollution it caused.
- Introduction of a pollution threshold. 500 pounds of waste or 10 pounds of toxic substances for companies. For towns 10 per cent of the waste. Companies which remain under the threshold will not be held responsible for soil pollution, financiers will not be touched. A basic fund will cover the costs for insolvent companies.
- The polluters responsible for the damage will not have to clean the site thoroughly and return it to its original state, but only to the level necessary for protecting public health and the environment.
- If at least 85 per cent of all companies involved in Superfund sites promise not to present claims to their insurance companies, the insurers will pay 8.1 billion dollars into a trust fund in a 1-year period. This fund will then cover the clean-up claims.

Discussions were facing difficulties in May 1994, particularly on the permissible level of remaining surface pollution.

While the Superfund deal now being negotiated is a step backwards for the environmental movement, the environment will benefit from the introduction of workable legislation. The cleaning-up of polluted sites is now being restrained by incessant court cases instigated by companies to avoid liability claims. Of the more than twelve hundred sites designated "Superfund sites" with serious surface pollution requiring an estimated 463 billion dollars by 2075, only 200 had been cleaned by early 1994.[21]

Another important impetus to the widespread application of preventative environmental management in the United States is legislation on the availability of information on corporate pollution, the so-called "Right to Know" law.

The "Right to Know" law in the United States

In 1986, the US Congress passed the "Emergency Planning and Community Right-to-Know" Act, or EPCRA. An important aspect of the EPCRA is prevention of chemical accidents and the promotion of disaster plans. Part of the law – the Toxics Release Inventory, also known as TRI, aims to inventize data on daily emissions of toxic chemicals into the air, water and land and of the flow of wastes to waste reprocessing plants. This national information bank has a uniform system of data input, and its output is publicly available.

While it is a relatively new programme, it has greatly influenced the way companies treat the environment, largely due to the public availability of the pollution data. Environmental organizations analyze the data and publish regular reports on companies and branches of industry. Everyone now knows how much pollution is caused by which company.

Once a step in the direction of preventative environmental management has been taken, the largest corporations don't hesitate to exploit it in their PR, advertising and marketing policies. The public is flooded with information on company environmental policy and products are promoted as "green" or "natural".

In general, this information and marketing activity aims to show how diligently the company is working to protect the environment, and too claim that the consumer helps protect the environment by buying its product. This is often a misleading sales tactic to improve the product's sales, and it clearly demonstrates that we can't leave the onus for supplying information on the environmental behaviour of transnational corporations to their own free initiative. By contrast, the US Right-to-Know legislation illustrates the positive effect of binding legislation.

The Green PR of transnationals[22]

"Environmental public relations" and "Environmental public affairs" offices of transnational corporations often communicate directly with individuals through informal discussions, claiming that this stimulates better understanding between the company and its target audience about its concern for the environment and its environmental activities.

For target groups where the contact is not as strong or groups which are difficult to reach with interpersonal communication, corporations use mass communication methods such as annual reports, leaflets, letters and press releases in newspapers. These aim at reaching shareholders, banks, consumers and (local) governments.

This is not a well-balanced relationship and the corporation always holds the advantage. It is one way traffic. Promoting mutual understanding and creating a positive environmental image has the purpose of improving sales figures, avoiding conflicts and influencing the target audience. In so doing the image of the corporation is protected and its future sales guaranteed.

A study carried out on behalf of the United Nations Development Programme by the organization Sustainability looked at the environmental reports of 100 corporations in 1992 and 1993. It was only able to identify five companies as coming "close to providing effective, useful data".[23]

Strategic environmental management

Corporations which implement preventative environmental management do as little as possible while maintaining effective operations with the greatest cost efficiency on the market and avoiding liability. Strategic environmental management places higher demands on the company, requiring an all round re-orientation in corporate planning, research & development and investment policy and the unambiguous dedication of top management as well as the integration of environmental management into all important activities.

An important tool in this type of company environmental management is the so-called *life cycle analysis (LCA)*. This is an analysis of the environmental problem throughout the entire production chain, including the waste phase of the product. This method is particularly useful in finding the environmental bottlenecks and risks.

Life cycle analysis

In 1985, the Swedish concern, Tetra Pak, commissioned a life cycle analysis (LCA) in order to establish priorities for environmental management. It was one of the first companies to use this instrument which is now being employed by an increasing number of corporations who claim to take environmental management seriously. The source of its popularity is found in several characteristics of LCA.

An LCA is an inventory of all raw and other materials, and all emissions and risks in all phases of manufacture and consumption of a product, including the waste stage. Using this overview, the corporate management can easily determine where measures should be implemented to reduce wastage of raw materials and reduce emissions and risks. The idea behind this is that every reduction in the use of raw materials and the amount of emissions and risks is to the benefit of the environment. Using this inventory, the management is not only able to establish priorities, for example concerning emissions, but, as with the green accounting system, it also gains an overview of the costs connected to these priorities. This opens the way for action throughout the entire chain on the basis of cost comparisons of possible measures.

But it has two distinct disadvantages. First, with an LCA it is not possible to compare the environmental "friendliness" of various products. If one product, for example, causes more CO_2 emissions while another requires more raw materials, what is the environmental criterium for choosing one above the other? Other data is needed to answer this question.

Secondly, LCA does not provide information on the environmental *effects* resulting from manufacturing and using a product. It only shows where emissions take place. Prioritizing based on the LCA does not take account of the severity and extent of the environmental effects and risks. This information can only be provided by undertaking an environmental effect and risk analysis of a specific situation, applying an "environmental value" to every emission and every risk. This sort of analysis will make it possible to base environmental measures on environmental problems. Nevertheless, to decide which product is environmentally "better" remains a subjective or "political" question, a calculation of "environmental values".

Apart from these two fundamental issues, there are also practical objections to the method, such as the fact that the continuing process of technological innovation makes the data become quickly obsolete, and regional and local circumstances have a strong influence on environmental effects (for example the presence of other sources of pollution which reduce the ability of the local ecosystem to absorb a certain emission).

Increasing numbers of transnational corporations use the LCA method. This is a step forward in comparison to green accounting because it doesn't simply look at *one* aspect of production – that of the company location – but at all the links in the entire production chain, including the consumption and waste phase of the product. In this way both, the problems and solutions at the beginning of the chain (prevention) and at the end of the chain (recycling) are visible *in relation to each other.*

The LCA method cannot, however, compare products. No single transnational will decide on the basis of the results of the LCA to stop producing a product, no matter how destructive that product and its production process has been. This is only an option if the sales of the product decrease (and it cannot, for any reason, be replaced by another product). This is the point at which corporate environmental management comes into conflict with the corporate means of production. Change will be precipitated only by external, political influences, in the form of environmental regulations.

This is the same point at which other instruments of the total environmental management conflict with corporate thinking. Dialogue with the public is applied to stimulate "mutual understanding", and external auditing and data is made available to increase public control over the corporation, increasing public trust. In themselves these represent a step in the right direction, but as long as competing corporations are not forced to alter their environmental behaviour, these measures will not lead to fundamental changes. And as long as the "availability of information" is not instituted, information will be distorted and suppressed and we will continue to see a series of one-sided success stories and a lot of patting of one's own shoulder.

The PR policy of Norsk Hydro in relation to the application of chemical fertilizer in agriculture

The Norwegian company, Norsk Hydro, provides an excellent example of how information is distorted for the purposes of publicizing the environmental aspects of chemical fertilizers. The largest manufacturer of such products in the world, Norsk Hydro also produces aluminium, magnesium, oil, gas, industrial gas and other products. It has production plants in virtually every part of the world, but most are in Europe.

The European chemical fertilizer market is shrinking rapidly. This is the result of EC cutbacks in agricultural subsidies, farmers' decreasing use of chemical fertilizers, and the increase in other forms of agriculture which use little or no chemical fertilizers. Norsk Hydro endorses the environmental need for careful application of fertilizers but contests the notion that alternative agricultural methods using animal and natural fertilizers is preferable, and has campaigned intensively on this theme for the past four years.

To avoid appearing biased, it commissioned an independent research into the environmental effects of both the alternative and the established methods of agriculture. The results have been published and widely distributed in seven languages (including Russian) in a very readable book. As could be expected, the established system of agriculture is portrayed most positively. Properly dosed, they say, chemical fertilizer is environmentally better than the manure used in alternative agriculture, since it is cleaner and cheaper for the consumer. This argument seems reasonable, but is not. Norsk Hydro limits environmental considerations in its book to application, not mentioning the preceding stages of production of chemical fertilizers. It does not mention that phosphate fertilizers are the product of a number of stages of manufacture in which "dirty" phosphate ore is cleaned and changed into chemical phosphate fertilizer. The significant environmental damage caused by this production process (for example through dumping cadmium and radioactive substances) are not visible after the product exists on the market. The price of chemical fertilizers does not reflect the cost of the environmental problems caused throughout the manufacturing process.

Neither of Norsk Hydro's conclusions, that chemical fertilizer is preferable to natural fertilizer for environmental reasons, and that the end product is less expensive to the consumer, are really proven. Essential information is omitted in this distorted argument that is being spread through all sorts of agricultural organizations in Europe and beyond.

In recent years, Norsk Hydro has been awarded several prizes for its environmental PR. We are left to wonder about the quality of the environmental PR of other transnational corporations.

Corporations employing strategic environmental management anticipate future rules and regulations in their research and development: the green R&D. They are active in consultations with civil servants, politicians and branch organizations to formulate new legislation and industrial codes of behaviour. In Northern European countries with a well-developed industrial state policy, the participation has shifted away from delaying or postponing the introduction of new environmental measures on a local or national level, to adapting these measures to the technical and economic possibilities available to the companies.

Lobbying for time

Hydro Agri Rotterdam is situated on the Nieuwe Waterweg in Vlaardingen (the Netherlands). It is a subsidiary of Norsk Hydro, the largest artificial fertilizer producer in the world. Directly opposite the plant, on the other side of a broad stretch of water, is another phosphate fertilizer producer, Kemira Pernis, subsidiary of the Finnish chemical giant, Kemira Oy. Both companies have been dumping highly contaminated phospho-gypsum into the Nieuwe Waterweg for years and environmental organizations have been at loggerheads with the two companies. The Ministry of Traffic and Water announced as early as 1988 that the amount of cadmium in the phospho-gypsum would gradually have to be reduced to nil.

Both companies negotiated with the local authorities for a delay. Hydro Agri Rotterdam announced it would rebuild two phosphoric acid units, not only to save energy, but also to insert an experimental cadmium purification technique into the production process. The company then applied the same delaying strategy used by Shell Pernis and described earlier. It was unclear at that point whether the experimental purification, which was still being researched, would work on a full-scale level. Research time was needed to find out. In 1992 the reconstruction work to achieve energy savings began.

On the other side of the water, Kemira Pernis embarked on a different strategy, declaring itself willing to invest in a cleaner installation. Time was needed. Not for research, but simply for the assurance that the government would not tighten its discharge restrictions for the next 20 years. Only under these circumstances would the company begin cleaning up its production technology, as this would ensure a guaranteed period of depreciation for the new installations, and not require new, far-reaching investments. If the government did not offer this security, Kemira Pernis threatened, it would close the phosphoric acid plant.

In 1991, it looked as though the environmental movement had achieved a victory. At a court hearing, the Council of State annulled the discharge permits of both concerns. The whole application procedure for new permits began again. The authorities permitted the companies to continue with their now "illegal discharges", granting a temporary permit until 1994. After this, the discharge standards were to be tightened even further.

In the meantime, one of the two phosphoric acid units at Hydro Agri Rotterdam has been closed down, as have the chemical fertilizer (pellet) plants at both Hydro Agri Rotterdam and Kemira. Hydro Agri has still not begun constructing its cadmium purification installation. Most of the employees have been sacked. Kemira had to reduce its workforce considerably, but the business is still in operation. Both companies now have new permits for the period up to 2000. Because part of the phosphoric acid production has closed down, Hydro Agri has no difficulties meeting the lower discharge standards for cadmium. Without large-scale investments, Kemira will have considerable difficulties with the new permit.

Sustainable development management

None of the forms of corporate environmental management mentioned so far can be labelled "sustainable". Only where a company regards its management policy as a series of extensive stages can we say that it is developing *toward* sustainable environmental management. This also applies to the fourth and last sort of environmental policy which distinguishes itself from strategic environmental management by explicitly taking international aspects of economic development and the global dimension of environmental problems into account.

According to UNCED, "environmental management directed toward sustainable development" should, in the *first* place, not only be globally applicable, but also make allowances for the specific circumstances of developing countries. This could be the development of specific policies and procedures for developing countries in the area of: training programmes, internal emissions/discharge standards (in the absence of government standards, or where existing standards are too weak), technological cooperation, interest in the local community and culture.

The Benchmark Survey found that only one or two corporations had referred to developing countries in their declared environmental aims. Moreover, only a few corporations stated that they applied the same environmental auditing standards throughout the world.

A *second* aspect of "environmental management directed toward sustainability" is the special attention given the role of the corporation in world-wide environmental problems such as the pollution of the oceans and atmosphere, loss of biodiversity and destruction of tropical rainforests. These are naturally coupled to international auditing and the public availability of data.

Once again, the Survey found that only a few corporations scored points for this, except in the gradual cessation of the production and use of CFCs which many corporations had undertaken. International agreements on banning CFCs (the Montreal protocol) seem to have had their impact on corporate policy. The authors concluded:

> "This suggests that when given a tangible and simple course of action the business community can respond with impressive positive environmental contributions, even in the case of rather diffuse and global environmental problems."[23]

A *third* aspect of "environmental management directed toward sustainability" logically follows on the preceding two: international cooperation on issues which fall into the area of contact between the business community and the environment. The Survey highlights the contribution of transnational corporations to the definition of international minimum environmental standards through active participation in their local business organizations.

If we look at the figures only, we can observe tremendous progress in international cooperation. The period leading up to the UNCED conference gave birth to the Business Council for Sustainable Development which represented member companies in the discussions on sustainable development. The International Chamber of Commerce also devotes a lot of attention to the discussion of the international environment. Thus, some 1,000 transnational corporations were involved in one way or another in the preparations for UNCED. The BCSD, which was originally established as a one-off coalition, is now a permanent entity. The international environment conference for industrial corporations, WICEM, is now a regular occurrence. Unfortunately, none of this has ever been aimed at developing minimum environmental standards or other forms of binding international regulations. As we saw in the first chapter, the aim of the corporations and their organizations was to prevent the development of such regulations.

In chapter 8 we will discuss the advantages and disadvantages of establishing international minimum environmental standards in relation to GATT.

Environmental management and business structures

The Benchmark Survey found that the success of company environmental policy depends heavily on the support and cooperation of the corporate head. If this is not forthcoming, all attempts at comprehensive environmental policy development are doomed. This conclusion seems diametrically opposed to the present wave of decentralization rolling through the international business community. Increasingly, larger companies are replacing their hierarchical, centralized structures with decentralized decision-making bodies, known as Business Units. In many conglomerates, the Business Unit is only required to provide corporate management with a financial report. This implies that the central management is unable to force environmental management policies on the affiliated companies in the Business Units. Not all corporations have this level of decentralization, and, in most cases, the corporate head bears final responsibility for finance and management. The exact content of this responsibility depends on agreements made between directors and affiliated companies. The decentralized application of centrally established environmental management policies, therefore, is not necessarily contradictory to decentralized organizational structure.

Summary

There are four types of environmental management in transnational corporations, ranging from "aimed at observing regulations" and "preventative", through to policies "toward sustainable development". A 1993 United Nations study of 210 transnational corporations with an annual turnover of more than 1 billion dollars, found that only a few had introduced policies aimed toward sustainable development. A closer look reveals that all four types of corporate environmental management have to be viewed with a degree of suspicion.

- In environmental management aimed at observing regulations, transnational corporations tend to negotiate with governing authorities for maximum manoeuvrability concerning production emissions. They introduce measures after the problem is caused, only to contain the damage to legally prescribed levels. Only a few transnational corporations have established corporate-wide policies aimed at keeping within the restrictions, leading to scepticism about the claim that transnationals will develop voluntary initiatives to maintain international environmental rules and regulations.
- It is not the recognition that environmental problems have also to be approached by business and industry, but the possible savings in costs and the financial risks of environmental pollution, that leads most large companies to introduce any form of preventive environmental management. Green accounting and internal audits for estimating and evaluating environmental costs are the instruments used, and this implies that the policy extends no further than the boundaries of the company itself.

  Once the step has been taken to introduce preventive environmental management most, and particularly the larger concerns, exploit these in their PR, marketing and advertising campaigns. A stream of industrial "environmental information" reaches the public, promoting the product as "green", "natural" etc. The often misleading and sometimes absurd nature of this sort of "environmental information" clearly illustrates that we cannot leave the supply of information on the environmental behaviour of transnational corporations to the voluntary initiatives of the companies themselves. Legislation concerning the availability of pollution statistics in the United States illustrates the positive effect of binding legislation in this area.

- Strategic environmental management requires an all-round re-orientation in company planning, research and development and investment, and the support of the corporate head. Life cycle analysis is a useful aid, as it estimates the environmental consequences of all stages of the production chain, including the waste stage, but it does not force even the most polluting companies to retire from business. At best, it leads to improving environmental management within the chain of production. This is also reflected in the PR policy of these corporations.
- Even the most far-reaching environmental policy "aimed at sustainable development", which only a handful of companies claim to practice, does not guarantee optimal results: world-wide application of the same environmental standards or auditing is only a step towards the application of the toughest standards. Paying attention to world-wide environmental problems is only useful if this leads to absolute reductions in particular emissions at company level.

All these limitations lead to the conclusion that more than voluntary initiative is needed to stimulate the evolution of corporate environmental management in the best direction from the (world) environmental perspective.,

CHAPTER 7

The inaccessible transnational corporation

Businesses fall under local and national, not international, laws and regulations. Since their activities are spread internationally, therefore, transnational corporations can exploit the differences in national rules, and they do so routinely in the case of labour legislation (regarding child labour and working hours, for instance) and taxation policy. The entire legal structure and internal financial organisation of these companies is based on fiscal differences, and Advisors on international fiscal law flourish on the specialism they euphemistically call "taxation planning".

The same situation applies to environmental legislation, where it is now possible for multinational enterprises to exploit differences in local legislation, free from the sanctions that could be imposed by international rules and regulations. A number of international organizations have been trying for years to fill this vacuum, and have proposed codes of conduct and internationally binding regulations based on agreements and conventions. Only in a few instances is this legislation specifically aimed at transnational corporations. This chapter looks at the degree to which they bind international companies.

Voluntary codes of conduct

The OECD and ILO codes, operational for several years, are specifically directed at transnational corporations as a separate group of entrepreneurs. They cover a broad range of subjects. In the OECD code, the "the polluter pays" principle has been operative since 1976, and, in 1992, it was expanded to include a section on the environment and accident prevention and a chapter on the responsibility of investors from OECD and non-OECD countries. For many years, the United Nations has been negotiating with representatives of the international business community to compose a code of conduct for transnational corporations. These were never completed and were finally transferred to the agenda of the transnational commission of the UNCTAD.*

Environmental codes of conduct

There are a number of other international codes of conduct created by international organizations which cover either a large number of subjects (including the environment), or one specific environmental issue.

General:

- OECD Guideline for Multinational Enterprises (OECD);
- ILO's Tripartite Declaration of Principles Concerning Multinational Enterprises (ILO-ME);

Specific:

- International Standards Organization's Technical Environmental Standards (ISO);
- UNEP's Environmental Guidelines (UNEP-EG);
- ILO's Code of Practice on Accident Prevention (ILO-AP);
- Conseil Europeen des Federations de l'Industrie Chimique's Guide to Safe Warehousing for the European Chemical Industry (CEFIC-SSC);
- CEFIC's principles and Guidelines for the Safe Transfer of Technology (CEFIC-TOT);
- FAO's International Code of Conduct on the Distribution and the Use of Pesticides (FAO);

---

* In the draft version of the UN code of conduct, where the negotiations have reached a stalemate, "environmental information", repair to damage to the environment and test practices for determining environmental safety are among the subjects discussed.

- UNEP's Awareness and Preparedness for Emergencies at the Local Level (UNEP-APELL);

There are also some more recent codes relating to the "environmental conduct" of transnational corporations which have been created by employers organizations:

- International Chamber of Commerce's Environmental Guidelines (ICC-EG);
- Chemical Manufacturers Association's Responsible CARE Program (CMA);
- The Business Charter for Sustainable Development;
- The Japanese Keidanren's Global Environmental Charter.

The last group are directed at the business and industry community in general rather than transnational corporations, which they mention only occasionally. The Benchmark Survey noted that in 1991 less than fifty per cent of the corporations in the study applied any of these codes to their practice. Later research indicated that one third of the companies in the Survey signed the 1992 Business Charter for Sustainable Development while seventy per cent of the Japanese companies had signed the Keidanren's Global Environmental Charter the same year.

International environmental principles?

The Japanese corporate organization, Keidanren, urges its members "to make environmental protection a priority at overseas sites ... apply Japanese standards concerning the management of harmful substances," and "actively work to implement effective and rational measures to conserve energy and other resources even when such environmental problems have not been fully elucidated by science".

The Business Charter on Sustainable Development of the International Chamber of Commerce (ICC) recommends corporations to apply the same company principles internationally and to watch that suppliers do likewise. The OECD recently developed "Guidelines for the Prevention of Accidents", advising that "hazardous installations in non-OECD countries should meet a level of safety equivalent to that of similar installations in OECD countries."

The international business community subscribes increasingly, it appears, to voluntary environmental principles. However, these codes of conduct do not require companies to apply the toughest environmental standards (as applied for example in the country of origin) internationally. Even the most far-reaching, that of the Japanese Keidanren only *recommends* the international application of local standards.

International application of environmental principles is, in all these codes, a very voluntary, individual decision.

Surprisingly little attention to the application of international company environmental management

The Benchmark Survey identified unambiguous results:[25]

"The participating companies were all large transnational corporations. Thus they could be expected to have extended procedures and policies for overseas subsidiaries and affiliates. However, both the statistical analysis and an evaluation of the material submitted by individual corporations indicated surprisingly little consideration for the international aspects of corporate activities. Regulatory discrepancies and the decentralized organization favoured by many TNCs may account for that finding.

Approximately half of the respondents had allocated Environmental Health and Safety responsibilities to their controlled affiliates; only 15 per cent had arrangements with their non-controlled affiliates. Other companies stated that they intended to observe local regulations. Some companies gave explicit accounts of their international responsibilities in their policy statements. Other corporations stated that they were prepared to establish their own standards if local ones were inadequate or absent. The more positive finding was that a handful of corporations had pledged to employ the same standards world-wide, thus meeting the recommendations of UNCED. That group included BF Goodrich, Amoco, AB Volvo, Union Carbide, Boehring, Ingelheim, and Ciba-Geigy."

The fact that active management concerning the environment, safety and health is restricted to the national level indicates the influence of national rules and regulations.

Conventions, guidelines and treaties

In recent years, one large environmental conference has followed the other, most with a specific goal. They resulted in resolutions, final declarations and protocols which can be interpreted as policy recommendations for the international community. The goal was sometimes more extensive, for example to establish a guideline or convention that a certain target group should observe, or in the hope that part or all of it would be taken up in national legislation. In these policy recommendations, transnational corporations as a specific type of company are markedly unmentioned.

United Nations environmental and health organizations such as the WHO and UNEP follow these guidelines, but, in general, they are applied on a voluntary basis and are weakened if they are not ratified world-wide. Nevertheless protocols and conventions can form the foundation of far-reaching international regulatory activities, which can directly influence transnationally operating corporations. To date, these are more the exception than the rule, the forced reduction of production of CFCs and the ban on trade in hazardous wastes being the best known examples, both of which were established following the path of the convention, but not without problems.

The Montreal Protocol

The Vienna Convention for the protection of the ozone layer (operational since September 1988) and the Montreal Protocol relating to substances which damage the ozone layer (operational since January 1989) created a policy resolution to gradually ban CFCs by 1996. Reacting to this, the world's largest producer of CFCs, Dupont, announced it would stop its production by 1994. This did not entirely solve the CFC problem, since not all the CFC product varieties fell under the original protocol, including

HCFCs, a CFC group with an extra hydrogen atom which is less destructive than other CFCs but does destroy ozone. Dupont and other producers have now switched to this HCFC to replace CFCs (for example in automobile air conditioning systems). In 1991, the Protocol's Scientific Assessment Panel concluded that CFC alternatives, especially HCFCs, were more harmful than expected. As a result, the Montreal Protocol was amended and this amendment ratified early in 1994 and became operational on June 14 1994. It was agreed that HCFCs and the halons will be phased out, but more slowly: by 2030.

The Basel Convention

The world currently produces around 400 billion tonnes of hazardous waste (including diluted waste water) per year, of which eighty five per cent comes from the United States and five to seven per cent from the European Union. The first agreement banning the international trade in hazardous (meaning toxic, radioactive or other health threatening wastes) was established in the Lomé Convention, which banned all transport of radioactive and hazardous wastes from the European Community to ACP (Africa, Caribbean, Pacific) countries. This was only applicable to Lomé Convention countries and left out 78 of those in the Third World.

The Basel Convention established a more detailed agreement, but with weaker content, rejecting a UNEP draft, and leaving only clauses on the "control" and "management" of the international trade, but no ban. This effectively legalized the trade in hazardous wastes under specific conditions, such as that known as Prior Informed Consent i.e. the receiving country must be informed beforehand and agree to the shipment. It is not difficult for a large, money-wielding company to acquire a signature indicating such consent. The Basel Convention also allows plenty of room for subjective interpretation and a resourceful trader will find it is full of holes. It has been so stripped of teeth that it can be easily bypassed, which occurs on a large scale.

Yet only 64 countries ratified this convention. In the meantime, an attempt has been made to reach a new agreement within the OECD banning the export of wastes from OECD countries. On March 25, 1994 the OECD countries agreed in Vienna to a total ban on exporting toxic wastes to non-OECD countries. The 64 signatories of the Basel Convention also signed this agreement, which, while not water tight – it has such escape routes as the category "recyclables" – is expected to cut off the new markets for toxic wastes which arose in Eastern Europe and Asia in recent years.

The most far-reaching efforts to create comprehensive multilateral agreements took place at international conferences, such as UNCED in 1992. But the treaties reached at such meetings have to be ratified by a certain number of the states whose governments participated before they become operational. Ratification means that the treaty is binding in that country, and it will abide by the rules of arbitrage and sanctions. The international effectiveness of a multilateral treaty is highly dependent on the number of ratifications. The more there are, the more it can be said that the treaty truly represents international legislation for our international legislation is fact based on consensus between states, and if there is no consensus the legislation does not exist.

What is wrong with international environmental law?*

"Based on these treaties, state practice and internal jurisprudence we can conclude that a number of principles in international law have been developed this century which form the foundation of modern international environmental law. The following principles fall under this category:

- the principle of sovereignty, that every state has the right to practice its own socio-economic and environmental policies.
- the principle of good neighbourliness, leading to the obligation to consider the interests of the neighbouring country and not to intervene unjustly and cause damage to its natural environment and economic activities and, in connection with this, the duty to provide immediate information and consultation, and

---

* In this passage we have quoted from "Advice on the environment: a world-wide problem. Towards a politic of sustainable development." Dutch National Advisory Board for Development Cooperation, No. 101 June 1993. page 40 ff.

- the principle of international liability for unlawful deeds. In certain circumstances this can be applied to across the border environmental damage.

In modern international law a number of important principles are being created. Besides the principles of good neighbourliness and international liability for illegal deeds, important in across the border environmental damage, these are in the main principles which flow from the duty of states to cooperate internationally and the involvement in, albeit the right of, people and nations to a healthy living environment.

Despite the positive developments, we have to conclude that the existing international body of legal instruments is inadequate. This is due to a number of factors which partly have to do with the as yet inept development of international environmental law and partly with fundamental characteristics of existing international law. The following problems can be seen:

- In international law the states are the most important actors. As a result not systematic legislature but disaster law and partial approaches are developed.
- All states, and particularly developing countries, meet with difficulties in translating international laws to creating national legislation and developing the controlling apparatus.
- Other actors, such as corporations (whether or not transnational) and individuals have almost no formal judicial status at the international level. It is difficult to charge them and to proceed against them on the international level for environmental crimes. Even at the national level experience indicates that it is often difficult for states, and in particular for developing countries, to successfully fight legal battles against international corporations.

It will not always be easy to coin development in the area of environmental law in a treaty text, either because the development of law is still in the growing stage or because important elements in it (specific duties and sanctions) are still being strongly fought against. The experience in other areas teaches us that in such cases the form of non-binding texts should be chosen. In most cases the text will to an extent become binding and will be felt in the development of other instruments."

Expanding body of environmental instruments

International environmental laws and regulations are patently limited. This is one of the reasons why international environmental policy has generally remained ad-hoc and reactive for a long time. The principles of Stockholm (1972) were an important breakthrough. But the first major, cohesive policy lines for the future were agreed to at UNCED, in its Agenda 21. Even here the transnational corporation disappeared from sight as separate category of business within the total business community, to which separate international treaties and conventions can be directed. No binding measures were formulated for this type of company. This was a success of the business lobby operating throughout the UNCED period (see chapter 2). Since UNCED, a great diversity of "instruments" have been developed which *national governments* apply in the context of national rules and regulations.

The diversity of the body of environmental instruments

There are a number of ways international treaties and conventions can be translated into national environmental policy. The UNCTC describes the following categories:[26]

1. Prohibition orders and legal instructions.
2. Demands relating to environmental effect analysis and planning for land use.
3. Demands relating to reporting, availability of information, monitoring and observation programmes.
4. Quality standards.
5. Demands concerning tests and availability.
6. Introduction of "the polluter pays" principle.
7. Introduction of the "prevention and precaution" principle.
8. Price and tax measures.
9. Subsidies and government investments.
10. Legislation pertaining to liability and compensation.
11. Executive and mandatory measures and dispute procedures.
12. Extra-territorial legislation.

These instruments have been developed for the application and use of those individual nation states that have ratified treaties and conventions. True international instruments, such as the saleability of internationally agreed national emission rights are still at the stage of opinionated discussion. National governments often apply different instruments in introducing or enforcing international treaties and conventions. This increases the existing diversity in the development of environmental policy in the various countries of the world.

Generally, transnational corporations find this diversity in rules and regulations difficult. The lobby for the harmonization of international environmental regulations was more important to international business and industry than the creation of environmental codes of conduct or treaties and conventions: 62 per cent of the transnationals researched in the Benchmark Survey want the United Nations to reform discrepancies in environmental rules and regulations. More than half are in favour of initiatives for the international development of a communal set of national environmental policies. The main aim of this harmonization is not so much simplification but the "levelling of the playfield" for all players on two levels:

- equalization of the conditions for *international* competition (between transnationals, that is to say, export industries)
- equalization of *national* conditions for competition (between national companies and importing companies).

This is the territory of *international trade relations*. At the present time, the most far-reaching development in international law is concentrated on this very territory: the GATT international trade agreement. The international business community has centred all available attention and influence on this event, as GATT does not only seem capable of simplifying and harmonizing legislation, but also weakening and nullifying strict national environmental rules and regulations. We will look at this more closely in the next chapter.

Summary

Despite the non-mandatory nature of specific codes of conduct pertaining to the way transnational corporations work, they have long fought the development of strict codes of conduct. In recent years, however, the international business community increasingly endorses voluntary environmental principles. But the existing codes of conduct are quite general, not demanding, for example, that companies apply the toughest environmental standards internationally. There is not monitoring and enforcement of these standards and no sanctions for companies who flout them.

Conventions, treaties and protocols are less voluntary and can form the foundation for far-reaching international regulations which directly influence the activities of the corporation operating in many countries. To date, however, these are exceptions, most notably the ban on CFCs and on trade in hazardous wastes. Both were devised – not without difficulties – through conventions.

Transnational corporations lobby against mandatory international legislation, and in particular against those agreements that identify them as a separate category. They also oppose the international diversity in rules and regulations. The international business community sees the lobby for harmonization in international environmental legislation as more important than the creation of environmental codes of conduct or international treaties and conventions.

CHAPTER 8

Trade and the environment: the transition from GATT to WTO

The Uruguay Round of the General Agreement on Tariffs and Trade (GATT), concluded in 1993, not only tightened the rules governing trade between nations but also set the tone for environmental policies of national governments. This chapter describes this connection and speculates on the future, when GATT/WTO will explicitly discuss the subject of Trade and Environment in 1995 and beyond.

From GATT to WTO

For fifty years, the GATT has been trying to develop a stable institutional framework for international trade by negotiating the reduction and abolition of measures which obstruct trade between member countries.

On the effect of the GATT agreements on global environmental pollution, opinion is divided between two camps. The first camp argues that the enormous economic growth which GATT will facilitate will be accompanied by an equal rise in world-wide environmental pollution. The second camp argues that economic growth will make funds available for environmental measures so that eventually global environmental pollution will decrease. Both agree that the economic changes the new GATT will cause in certain branches of industry will also bring changes in pollution. Opinions vary as to the significance of these changes for the environment.

The World Trade Organization (WTO), the 1995 successor to the GATT and executor of the new GATT agreement, will have increased authority to approach players who do not observe the rules. The WTO will also become involved in areas not yet clearly delineated including environmental pollution standards and possibly also social law (for example standards relating to child labour). It is still not clear what the outcome will be of the environmental standards, but the contours are being defined.

Dissatisfaction with the GATT concerning environmental standards

"Environment" is a reasonably new subject in the discussions on world trade. Until 1992, the GATT working group "Environmental Measures and International Trade" lay dormant. The discussion was set in motion after a GATT panel made a judgement in the Tuna-Dolphin case, and after the UNCED environmental summit in Rio de Janeiro. Since then, the environmental working group has worked intensively on preparing an agreement on the treatment of environmental measures within the WTO.

The Tuna-Dolphin case

This concerned a dispute between Mexico and the United States on the import ban on tuna in the US based on the Marine Mammal Protection Act of 1972, the Dolphin Protection Consumer Information Act of 1990 and the Fisherman's Protective Act of 1967. US fishing boats were given permission by the US to catch yellow tuna in the eastern part of the Pacific Ocean, on condition that the catch include no more than 20,500 dolphins per year. It was forbidden to import of tuna from fishers and countries that do not uphold the conditions of the aforementioned legislation.

In January 1991 Mexico requested GATT to adjudicate this matter. A GATT panel reached its decision in the dispute on September 16th 1991. The US import ban was forbidden. The two most important considerations were:

– the ex-territorial nature of US legislation in this issue: the US could only

ban the import of products if these products would cause health problems to people, plants or animals in the United States itself.
- the US ban on imports was based on the nature of the (catching) procedure and not on the characteristics of the product: that dolphins are accidentally caught when fishermen fish for tuna does not influence the "product" tuna, according to the GATT panel. Only imported products which have negative effects can be excluded from the market.
- The United States tuna embargo is also forbidden because it places US fishers in an advantageous position.

The tuna-dolphin decision illustrates that the present GATT rules make a strict delineation between import restrictions based on negative environmental effects the *product* will cause in the importing country (in that case import restrictions are allowed, in principle), and the restrictions to import of products which *were produced under circumstances which created environmental damage.* This second type is also known as trade restrictions based on the environmental *production standards* (this trade restriction is *not* allowed).

GATT and the environment

According to present international rules established in the GATT, a governmental measure to restrict trade, based on environmental considerations, must conform to extremely strict conditions.
- A measure may not be directed toward a single country, while the import of similar "dirty" goods from other countries is left unhindered.
- No distinction may be made between foreign and national producers. This principle prohibits measures which restrict trade in imported products with the purpose of protecting national business and industry (with the exception of measures to protect the exhaustion of national resources).
- A country has to prove that the trade restrictive measure is

necessary and effective in protecting the own environment. The GATT will prohibit the measure if it is not backed by scientific proof. The onus of proof rests with the country employing the environmental policy.
- If there are alternative policy measures which do not conflict with the principles of free trade, these must be used.
- Finally, a country may not interfere in the environmental problems in other countries, if those problems have no effect on that country. A country may not introduce import prohibitions or levies against products in which the production process caused pollution or in which prohibited substances (such as CFCs) were used. If the product contains prohibited substances or pollutes the environment, import may be prohibited.

These restrictions to measures that countries employ to restrict trade using environmental arguments will be increased in the WTO. The unambiguous starting point is that international trade must function with as few restrictions as possible. This will be the principle from which international negotiations within the WTO will take place, on the conditions, including environmental production process standards which may form a legitimation for trade restrictions.

Harmonized international environmental standards and scientific verification

What will the WTO's new environmental agreement look like? Under which conditions will countries be able to preclude the import of goods in which the production process has caused pollution? Considering the increasing unemployment in many industrialized countries it will be easier to find political support for import levies on "dirty" products than for decreasing their own environmental standards or mass lay-offs. Will the WTO allow this?

The GATT's Trade Negotiations Committee released a decision on December 15, 1993 on the starting points and the subjects of, among others, it agenda for trade and the environment.[28]

The working agenda of the Trade Negotiations Committee concerns:

– the identification of the relations between trade measures and environmental measures directed toward promoting sustainable development.

– developing recommendations possibly required if the Multilateral Trade System is to be adapted, in particular concerning:

- rules relating to trade and environmental measures (with special attention for developing countries) aimed at sustainable development.
- avoiding protectionist trade measures, and maintaining an effective multilateral discipline to ensure that the Multilateral Trade System reacts to environmental goals, including principle 12 of the Rio declaration, and
- supervision of trade measures with an environmental aim, and of trade related aspects of environmental measures which have an important influence on trade, and to an effective introduction of the multinational discipline concerning these measures.

The discussion has not yet crystallized. The preparatory working group Environmental Measures and International Trade is still acquainting itself with the possible discrepancies between GATT principles and national environmental measures. The negotiations are still to begin and it is not clear where they will lead. The new GATT rules, which are the result of the Uruguay Round discussions on safety and health standards, do however indicate the sorts of discussions and problems we can expect concerning environmental production standards.

Sanitary and phytosanitary measures

This is the heading under which the new GATT will combat trade restrictions imposed by governments with what they call "unfounded" measures concerning the safety of goods for the consumer and the prevention of the spread of animal and plant diseases. The new GATT does not want to prohibit these sorts of measures, even if they restrict trade, but to make them "transparent", and to prevent arbitrary measures. The GATT will encourage governments to base this sort of measure on international standards and codes of behaviour.

The GATT will only allow a country to apply stricter production standards than the internationally accepted standards if the government of that country can support the measure with scientific evidence or with a risk analysis.

These new GATT rules have far-reaching consequences. For example, the fact that governments may only base their import restriction measures on scientific evidence and risk analysis excludes the possibility of restricting imports on non-scientific grounds, such as religion, culture and ethics.

The example of the campaign against Nestlés instant milk powder for babies, and against other manufacturers of milk powder illustrates this. The milk powder itself was scientifically proven to be of good quality, but many of the babies in Third World countries who drank this milk died. They lived in areas which lacked clean water or the facilities to boil the water. In addition, powdered milk advertisements changed the traditional preference for breast feeding to a preference for instant powdered milk. Under the new GATT rules, governments in developing countries will no longer be able to prohibit the import of instant milk powder for babies.

Because the GATT will only recognize production standards on scientific grounds, the World Health Organization's (WHO) recognized production safety and risk standards are no longer valid. The WHO recognizes that economic and other standards are important aspects of the acceptance of a product in a country. This can be seen in the code developed by the WHO in cooperation with UNICEF against marketing products which replace breast feeding. The new GATT now explicitly refers to two institutions for developing and standardizing internationally recognized scientific product standards: the FAO "Codex Alimentarius" and the "International Office of Epizootics". Without wishing to cast doubt on the quality of the text of the Codex Alimentarius, it is striking that various sources have accused these two institutions of being strongly influenced by multinational chemical and food corporations.[29]

Analogous to the example of "Sanitary and Phytosanitary Measures", we can expect a strong trend within the Commission for Environmental Measures and International Trade to argue for international harmonization and scientific grounds for pollution standards for manufacturing *processes* (just as safety and environmental standards for the *product* required a scientific basis).

With this approach, it will only be possible under certain circumstances for countries to prohibit the import of goods, or impose levies, if these goods are produced under standards less strict than their own. As with product standards, the country wishing to place import restrictions will have to prove scientifically that their own strict process standards are necessary for the protection of their national environment.

From a global environmental perspective, there are two important points to be made on future GATT/WTO rules which will be based on internationally standardized process-pollution standards. The first is a criticism of the content of the approach, the second highlights its possible undesirable environmental effects:

The *first criticism* indicates that there is consensus in the literature that international harmonization on environmental standards is unnecessary and even undesirable. Countries have different natural resources, environments and different environmental capacities and every country has to balance its environmental and income issues. In formulating the scientific basis for more stringent pollution standards, it is important to take these aspects into consideration. This is no easy task.

Even if we limit the scientific evidence to purely ecological factors, it would be highly contentious for two reasons.

– Ecological capacity at the local and national level, as we saw in chapter 5, is not only influenced by *one* sort of dangerous emission (or risk) or by *one* source of pollution, but by a combination of emissions, sources of pollution and risks in that area. Progressive

government policies which recognize the cumulative effect apply stricter pollution standards to specific production processes: in area with a lot of polluting industries, the pollution standards per company will be stricter. International harmonization of production pollution standards negates this.
- The scientific approach provides no single, unambiguous starting point on which to base standards. It offers, at best, indications. It can never be proven *exactly* how much damage is caused by an emission. And scientific debate is still raging on the degree to which CFCs damage the ozone layer.

The *second criticism* of international harmonization and application of production pollution standards in relation to trade restrictions concerns the effect on the environment.

Internationally defined production and product pollution standards can lead to a decrease in global pollution if these are minimum standards and if the minimum standards are not too low, since it will be possible to force countries that have no interest in the environment, or that export environmentally harmful goods, to produce according to the minimum standards on pain of exclusion from world trade. This is why the environmental movement has been campaigning for many years for minimum standards.

The danger is that these minimum standards will be treated as maximum standards, in which case they will be too low. If a country which has above-average pollution standards tries exerting import restrictions on "dirtily produced" products, external sources may put pressure on those standards. The threat of having to provide scientific proof for its more critical process-standards might discourage such a country from applying trade restrictions.

This criticism of the future GATT/WTO rules is from the perspective of the global environment. There are others from the perspective of the economic development of developing countries.

Non-governmental organizations have been campaigning for years for the introduction of global minimum standards in the interest of the environment, social justice and health. The goal was to protect the interests of people and the environment in underdeveloped countries where legal standards were lower or did not exist at all. Representatives of (international) employers organizations, transnational corporations and governments in the developed North have steadily resisted this, arguing that such standards would endanger the stability of the world market and be an unacceptable violation of the sovereignty of national governments. Southern governments which lagged behind in establishing and applying national environmental management (or legislation concerning social justice) easily joined this campaign. Now that Northern countries increasingly want to influence environmental management outside their own national boundaries, and the harmonization of environmental standards is threatened, the standards in the North will not only, in all probability, be levelled down; in countries which have as yet few regulations concerning the environment and social justice the standards will be levelled up. There is a dichotomy of interests between the North and the South. The Southern countries now fear trade restrictions on their products as these are being produced under lower environmental standards than the likely world standard. This is unacceptable for these countries as it is the rich countries that are most responsible for the worldwide environmental problem. The problem is clearly outlined in the following quotes:

Green protectionism

"Trade measures are seldom the best means of approaching an environmental problem, is the opinion of the Dutch government. It is better to get to the core of the problem. It is preferable to transfer technology and money, to positively stimulate more environmentally friendly production and to participate in international consultations. The government does not however exclude the possibility of applying trade sanctions, as a last resort, to punish uncooperative countries, particularly if the interests of the international environment are at stake.

Because it is the rich countries that contribute most to the global environmental problem, it is not right to pass the buck to the poorer countries via trade. The Dutch government agrees with this standpoint. Ideally sanctions against a certain product are compensated through the transfer of technology or money to the supplier. Tragically, in the period since the UNCED environmental conference, the richer countries have kept their purse strings closed. This makes it even more difficult for the poor countries to accept "green protectionism". They will do whatever is in their power to fight environmental demands – or social demands – which damage their competitive position on the market being forced upon them."[30]

Social protectionism

France and the United States are in favour of the new World Trade Organization (WTO) exploring the issue of social justice, in particular the desirability of developing minimum rules on which trade restrictions can be based. The Third World is unequivocally opposed to this "hidden protectionism" of the rich North, saying it would cut the low-wage countries short and protect their own industry. The ostensible issue is child labour and forced labour.

The European Union is, as usual, divided on this issue. The Netherlands belongs to the camp opposed to the French and US proposals, arguing that we ought not weigh the WTO down with non-trade, sensitive political issues. Belgium supports France. The Minister for Foreign Affairs R. Urbain proposed in the EU context to provide extra trade advantages to countries with a "good social report".[31]

There is less likelihood that the GATT/WTO discussion will end in a stalemate between Northern and Southern standpoints as happened at UNCED. Rich trading nations dominate negotiations within the GATT/WTO. Moreover each country is likely to be given ample time to grow toward the international harmonized minimum process standards. Whether the global environment will benefit from this depends heavily on the ability of the other countries to maintain and improve their own environmental management.

The position of transnational corporations

If a new GATT/WTO agreement on trade restrictions based on process standards resembles that on product standards, the transnationals will profit most. They can adapt their production installations in the South to the relatively weak international pollution standards, which removes any threat of trade restrictions on their own products. They can then lobby governments in the North to apply import restrictions to "dirtily produced" products from the South and encourage Southern governments to demand a GATT/WTO panel decision on Northern trade restrictions. In this way they can manipulate the GATT/WTO to open the borders in the North and to destroy internal process standards.

Transnational corporations do not need to create new lobbying and negotiating fora for GATT/WTO as they did with UNCED. The path has already been created and there are already openings and lobbying channels to the most important governments.

The influence of the US company lobby concerning trade

In May 1990, when it looked as though GATT would fail due to the differences between Europe and the United States, a number of leading companies in the US set up the Multilateral Trade Negotiation Coalition (MTN). Companies strongly dependent on international trade such as American Express, General Motors, IBM and Cargill were among the participants. The goal of the coalition was to influence the GATT through determined lobbying. Its success was clear during the July (1990) meeting of the G7, when, after a briefing with the MTN, the Bush government gave the Uruguay round top priority.[32]

There are many other examples of the power of the US company lobby. For example, the US cigarette market declined in the 1980s, while markets in other parts of the world were protected against imports either through monopolies or high imports tariffs. The US government took several South-East Asian governments in hand, threatening unilateral trade sanctions against Japan, South Korea, Taiwan and Thailand. Those markets were eventually opened, and this led to a tremendous growth in American exports of cigarettes from the large tobacco multinationals (Philip Morris, Reynolds), to the detriment of local producers.[33] The pharmaceutical industry provides another example. The WHO worked for many years to compile a list of essential medicines which Bangladesh took seriously and banned all unnecessary medicines. The US government threatened to cut food aid if Bangladesh continued to discriminate in this way against the US pharmaceutical industry.[34]

Summary

The environment is a relatively new subject in discussions on international trade. To date, import restrictions on products produced under environmentally hazardous conditions are not allowed under the GATT treaty. But Northern countries are imposing more and more demands on the quality of products (product standards) and on the way in which they are manufactured (process standards). Countries want to be able to refuse products which have dangerous properties and, increasingly, also, those for which the manufacturing process created environmental hazards.

The new GATT has bound import restrictions on products which are dangerous to people and the environment to very strict rules. Within the WTO – the successor to GATT – a working group is currently preparing agreements which make it equally possible to impose, in specific circumstances and according to specific rules, import restrictions on products manufactured under dirty conditions. These new rules may resemble the rules for refusing a product with hazardous properties. We can expect that the new rules will be based on internationally-agreed pollution standards, and that a product can be refused if it was manufactured under sub-standard conditions. If the manufacturing conditions of a potential import product are, in terms of the environment, lower than a country's own norms but not lower than the agreed minimum, a country will probably not be able to refuse it on environmental grounds unless it can scientifically prove that the country's own stricter production emission standards are necessary to protect its own environment.

Internationally applicable minimum process emission standards can have a positive effect on countries where standards are lower, as these countries will be threatened with trade restrictions. These, however, can have serious consequences for economic development. Countries should be given ample time to be able to meet the minimum standards. Transnational corporations can lobby for trade restrictions on the import of goods produced under polluting circumstances and can at the same time push for a GATT-panel decision against such restrictions, which will effectively put pressure on the stricter standards in the North.

CHAPTER 9

Conclusion

The central question of this book is whether international environmental management can be left to the voluntary activities of transnational corporations. Multinationals themselves have given their own answer: they are opposed to international, binding rules which identify them as a separate category of company and throw up restrictions. There are however several strong arguments against leaving the initiative to the transnationals. We can summarize these under five headings:

- the large measure of involvement transnational corporations have in, and the even larger responsibility they have for, environmental pollution
- the possibilities available to transnationals for avoiding stricter environmental legislation by moving operations elsewhere
- the imperfections in the development and dissemination of their environmental technologies
- the general inadequacy of international corporate environmental management
- the tendency to support only generally formulated and non-binding codes of behaviour while there are several examples of the effectiveness of binding international legislation.

This calls for international, binding environmental rules for transnational corporations as a separate category within the business community. However, it is too much to expect of the international political community that this will happen. The far-reaching international regulations presently applicable to trade – the new GATT and the future WTO – will not have a positive influence on the environmental behaviour of transnationals.

APPENDIX I

GATT and the patenting of life forms[35]

Not only French farmers strongly objected to the treaty text. In India there have been large scale protests against the new GATT. Indian farmers, scientists and environmental activists object to the chapter on Intellectual Property Rights, or patenting and licensing rights. Shortly before GATT was signed, on April 5th 1994, 200,000 people demonstrated in New Delhi against the treaty.

The issue is one in which the Indian farmers and environmental activists have taken the initiative, but the repercussions will be felt in all Third World countries.

The new GATT treaty prescribes that all signatory countries accept the patent laws of the Western industrialized countries. Everyone wishing to use a patent which has been registered in the West will have to pay royalties – even if the royalties are unreasonably high or have been abusively obtained. This is what upset the Indian demonstrator's.

Copyright rights are not accepted in many Third World countries at present, but, on pain of being excluded from the GATT systems this will have to change. The Indian Patent Law of 1970 forbids in principle individuals or companies patenting any form of life (plants, animals, seeds). Many other Third World countries have a similar law.

Since the 1960s, transnational corporations have applied in their country of origin (Western) for patents on numerous crops, seeds and active parts of plants and trees in the Third World. The American professor Iltis, for example, discovered two wild varieties of tomato in Peru which could be cultivated in the United States with good results. He imported the seed, took out a copyright and sold the copyright to a food company. There's not a pizza sold in the US without this variety of tomato in its sauce. Most patents are not however used in food production. The American Type Culture Collection, a scientific registration organization in Maryland Virginia, houses 60,000 patented or patent-ready organisms. Most of these are patents on parts of plants and seeds which are used in the production of environmentally friendly fertilizers, insecticides and medicines – for which there is a fast growing market in the West and with which companies are earning more and more money.

When the new GATT copyright law is introduced, US patenting laws will be applicable world-wide. Companies and farmers in the Third World, which have used these seeds and live on these crops for years, will suddenly have to pay for using them if the transnational copyright owners claim their rights.

APPENDIX 2

The Global Environmental Facility (GEF)[36]

The GEF provides financial grants to developing countries. Through three executive organizations – the World Bank, UNEP* and the UNDP – the GEF grants money to projects which have been set up to participate in some way to the protection of the world's global environment. The projects also have to conform to development goals. In principle, the GEF only finances incremental costs which are accrued in the interests of the world-wide environment, additional to the costs a country itself is expected to finance for its own nature and environment. GEF projects relate to the greenhouse effect, the pollution of international waters, the loss of biodiversity and the destruction of the ozone layer.

The GEF was set up in 1990, before UNCED, as a three-year experiment. The pilot phase was completed at the end of 1993. Since then, the GEF has been named as the interim financing mechanism for the UNCED Conventions relating to biodiversity and climate changes. After 16 months of negotiations the representatives of more than 80 countries came to an agreement on March 16, 1994. In the period 1994-1996, a number of Northern countries will donate 2 billion dollars to the fund.

NGO criticism of the GEF operations in the first three probation years is extensive, concentrating on two main themes:

- The first is directed toward the discrepancy between local environmental problems related to poverty and the international

---

* UNEP: United Nations Environmental Programme. UNDP: United Nations Development Programme

environmental problems which originate primarily in the rich world. The GEF is the only channel into which extra funds are pumped, but the GEF concentrates on the global environment and not on the daily needs of the poor countries. According to the critics, the value of the GEF is extremely limited, and, if it continues to refuse to address the major problems of poverty, it will have the opposite effect to what it is trying to achieve. The participation of the World Bank in GEF projects, the critics say, leads to a lack of answerability to the local population, and stands in the way of the reasonable participation of the local participation in the identification, development and execution of projects.

– A second point of criticism is directed toward the wrongful use of the funding. Critics say that developing countries apply for GEF grants for so-called incremental costs for global environmental protection while the local and national environment is inadequately insured, or the projects do not contribute to the development of the countries. If the country does not have adequate finances for a project, it will do all it can to find GEF funding. As a result, the GEF is confronted by projects which have no single connection with other projects and which are usually initiated and installed by outsiders. This does not strengthen the country's institutional capacity for sustainable development or its ability to protect nature and the environmental.

Nigeria, GEF helps Shell?

Nature and people suffer the serious consequences of oil exploitation in Nigeria. Pollution, noise pollution and the infrastructure needed for oil extraction, have paid their toll in the vulnerable mangrove forests, internationally recognized as a threatened ecosystem, and the provider of the livelihoods of thousands of people. All efforts of the local population to protest against the pollution and the oil extraction by which they barely profit are hard handedly repressed. On April 30 and 4 May 1994 one person was killed and many were injured during a demonstration against the installation of a pipeline by Shell, which is very active in this region. Ken Saro

Wiwa, one of the leaders of the Ogoni (an indigenous people in this region) resistance was arrested by the Nigerian authorities. The reason for his arrest was the international attention he had managed to generate for the problems of oil extraction in this area. The global climactic effect of the enormous amount of gas released in oil extraction led to the decision of the GEF to invest in the area. There is no mention in the plans, however, of the local environmental problems and there is no interest whatsoever in the problems of the local community.

The NGOs conclude that the principle of 'incremental costs' can only be defended if there are other channels available for working effectively on local and national sustainable development. Finance ministers will stop being willing to finance a GEF which continues to be wrongfully used. The NGOs also felt the GEF should be used to intensify the effect of the regular assistance channels. The GEF should certainly not serve to aggravate poverty situations, and should preferably contribute to the improvement of the local social situation.

NGOs also felt that the World Bank, which presently manages the GEF, should provide more and faster information on the candidate projects, and the concerned populations and NGOs should be able to become more involved in the decision-making procedures. Key concepts for a new GEF are transparency, accountability and participation.

The new GEF partially meets these points of criticism.

The executive structure is no longer so closely linked to the World Bank. A board of directors from the 32 founding countries (18 donor countries and 14 recipient countries) will be established. Openness, consultation and participation of the relevant NGOs and local populations throughout the entire project cycle has been taken up in the agreement. The participation of NGOs in the Board of Directors has not, as yet, been agreed to.

It is expected that the GEF will decide on its first work programme in December 1994.

References

1. "Uitstoot kooldioxide moet ver onder bestaand niveau". *Volkskrant* 16 September 1994, p.8
2. "Zorgen voor morgen" RIVM, 1988, Bilthoven, the Netherlands
3. ECN: *Environmental Review.* July 1992, p.8
4. World Development Movement, "Costing the Earth". 1991, p.21
5. "Green Rights for all: the earth view." Aubry Meijer. Global Commons Institute. In *ECN-Environmental Review.* July 1992
6. In: *Crosscurrents,* No. 4. 19 August 1991, p.4
7. The Malaysian ambassador Razali Ismail, leader of the delegation from his country. In *Onze Wereld,* October 1991
8. "The corporate capture of the earth summit" Benno Bruno, *Multinational Monitor.* July/August 1992, p.6
9. "The Business Council for Sustainable Development – Phase Two?" The Network, No. 21 November 1992, p.16
10. "Ongoing and future research: transnational corporations and issues relating to the environment." UNCTC, October 1989, p.5
11. "Transnational corporations and the issues relating to the environment." Report of the Secretary-General. UN Commission on Transnational Corporations. 11 January 1990.
12. UNCTC 1988, p.230
13. "The dye is cast by growth and costs. European chemical companies are shifting bulk capacity to Asia." Paul Abrahams in *Financial Times,* May 31 1994
14. Benchmark Survey, p.2
15. Benchmark Survey, p.65
16. By: FLS Miljo Environmental Management, a company in the Danish F.L. Smidth Group, in cooperation with the Danish Ministry for the Environment and a fertilizer producer. *Phosphorus & Potassium,* No. 162, July/August 1989, p.28

17. *Phosphorus & Potassium*, No. 148, March/April 1987, p.12
18. "Environmental Management in Transnational Corporation. Report on the Benchmark Corporate Environmental Survey." United Nations Conference on Trade and Development Programme on Transnational Corporations. Environmental Series No. 4, UN, New York 1993
19. "Rapportage over mogelijkheden van onderzoek naar milieuinvesteringen bij Shell Nederland Chemie. Hans Heerings, SOMO, Amsterdam 27 November 1989. Commissioned by Greenpeace the Netherlands. p.18
20. Benchmark Survey, p.169
21. "Toxic Turnabout? a deal on Superfund may finally be at hand." *Business Week*, 25 April 1994, pp. 34-35
22. From: "Groene markten", SOMO, 1994
23. "Dow best of bad bunch concludes UNEP report". *European Chemical News*, 11 July 1994, p.30
24. Benchmark Survey, p.176
25. Ibid p.65
26. "Emerging Trends in the Development of International Environmental Law at the Regional and Global Level: Implications for Tansnational Corporations." Prepublication Advance Unedited Copy. United Center on Transnational Corporations, UN, New York, 1992
27. *Focus.* GATT newsletter. August/September 1993, pp. 6-7
28. "Trade and environment". Decision, adopted 15 December 1993 by the Trade Negotiations Committee
29. "Business regulation and competition policy. The case for international action." Harris Gleckman, Riva Krut, *Christian Aid*, July 1994: "A body stocked with industry handmaidens from the USDA and FDA, wide open to lobbying by (...) transnational corporations." p.19
"The General Agreement on Tariffs and Trade, Environmenal Protection and Sustainable Development", C. Arden-Clark. A WWF international discussion paper. *Revisited,* November 1991.
"The Codex Alimentarius panel is known to be heavily influenced by multinational food and chemical companies. To

take but one example, there are 28 members of the US delegation to Codex Alimentarius, 16 of whom represent food and agrochemical companies or producer associations." p.28

30. "Wereldhandel staat voor 'groene' horde. Ontwikkelingslanden vrezen milieunormen rijke landen." Jolke Oppewal, *Internationale Samenwerking*, May 1994, p.35
31. *Het Parool*, 13 April 1994
32. *The Ecologist*, Vol. 20, No. 6, Nov/Dec 1990
33. "Philip Morris. Ontwikkeling in de divisie tabak in Europa en in de Verenigde Staten." Hans Heerings, SOMO, Amsterdam 5 April 1994
34. *The Ecologist*, Vol. 18, No. 2, 1988
35. Taken from: "Werelddiefstal", Aart Brouwer, *Milieudefensie* 4-1994, pp. 12-13
36. "$2 billion for Rio Follow-Up" *The Network*, No. 36, April 1994, and "Governments Push Ahead with the GEF." No. 24, March 1993. *E&O* No. 3: October 1993